Shaping a Digital World

Faith, Culture and Computer Technology

DEREK C. SCHUURMAN

IVP Academic

An imprint of InterVarsity Press
Downers Grove, Illinois

InterVarsity Press
P.O. Box 1400, Downers Grove, IL 60515-1426
World Wide Web: www.ivpress.com
Email: email@ivpress.com

InterVarsity Press® is the book-publishing division of InterVarsity Christian Fellowship/USA®, a movement of students and faculty active on campus at hundreds of universities, colleges and schools of nursing in the United States of America, and a member movement of the International Fellowship of Evangelical Students. For information about local and regional activities, write Public Relations Dept., InterVarsity Christian Fellowship/USA, 6400 Schroeder Rd., P.O. Box 7895, Madison, WI 53707-7895, or visit the IVCF website at <www.intervarsity.org>.

While all stories in this book are true, some names and identifying information in this book have been changed to protect the privacy of the individuals involved.

Cover design: Cindy Kiple
Interior design: Beth Hagenberg

Images: Web icons: © Ryan Putnam/iStockphoto
Abstract squares: © Pavel Khorenyan/iStockphoto
Mobico icons: © O'Luk/iStockphoto

ISBN 978-0-8308-2713-8 (print)
ISBN 978-0-8308-8444-5 (digital)

Printed in the United States of America ∞

Library of Congress Cataloging-in-Publication Data

Schuurman, Derek C., 1967-
  Shaping a digital world : faith, culture and computer technology / Derek C. Schuurman.
    pages cm
  Includes bibliographical references and index.
  ISBN 978-0-8308-2713-8 (pbk. : alk. paper)
  1. Computers—Religious aspects—Christianity. 2. Technology—Religious aspects—Christianity. I. Title.
BR115.C65S38 2013

261.5'6—dc23

P	21	20	19	18	17	16	15	14	13	12	11	10	9	8	7	6	5	4	3	2	1
Y	31	30	29	28	27	26	25	24	23	22	21	20	19	18	17	16	15	14	13		

Contents

Preface

This book began as a loose set of notes that I collected in an attempt to answer the following question: *What does my faith have to do with my work as an electrical engineer?* I was familiar with the notion that all of life falls under the lordship of Jesus Christ and that we can serve him equally well as a minister or a webmaster. However, when I found myself sitting in a cubicle farm and busily working in the high technology industry, it was increasingly difficult to determine exactly what impact my day-to-day work had in the kingdom of God. It's easy to say that faith informs all of life, but that notion becomes little more than a platitude without a more detailed understanding of the phrase.

I was educated in electrical engineering, which is to say that I was not well-educated in anything else. I received an excellent technical education, and upon graduation I felt confident I could tackle whatever technical challenges would come my way. It soon became apparent, however, that my excellent technical training had not provided me with a context for my work. More specifically, it was not clear to me how my faith related to my work.

I began to read and think about this question, and it persisted over the following years as I left work to pursue graduate studies in engineering. Eventually, I felt the call to move from industry into the area of teaching. I am thankful to God that he led me to a Christian academic community in which the question of how to integrate faith and learning is taken seriously. Whether you are in industry or studying in a secular or Christian setting, it is important to grapple with the call to "take captive every thought to make it obedient to Christ" (2 Corinthians 10:5).

The Bible tells us that God has chosen the time and place in which we live (see Acts 17:26). I am grateful that I was born in such exciting times! The first computer on a chip was invented a few years after I was born, and as I grew up I witnessed the introduction of the first personal computers, the development of the Internet and many other exciting digital technologies that have shaped the world. My teenage years were spent playing with electronic projects, exploring ham radio and learning how to program some early personal computers. Later in life, as a professional working in industry, I enjoyed designing electronics and writing software for some "real-world" applications.

This book is an attempt to provide both practitioners and students working in fields related to computer technology a beginning framework for discovering how their faith relates to their technical work. Many of the ideas in this book are not novel—borrowing a phrase from Donald Knuth, I would say that when it comes to philosophy and theology, "I'm a user, not a developer."[1] Foundational work in many disciplines is often the work of amateurs: those who are immersed in a particular discipline are rarely experts in philosophy or theology, and likewise, experts in philosophy and theology are rarely experts in another discipline. This should not discourage us, however, from the work of humbly forging a Christian perspective in our given vocations. This book only sketches the outline of a Christian perspective, and much hard work remains to address in more detail the implications of a Christian worldview for the many issues that arise in computer technology. My hope is that this book will provide a helpful contribution to the ongoing dialogue about faith and computer technology and that it will help spur further work in this important field.

I am thankful to stand on the shoulders of many others, and I owe much of what I have learned to the authors of the books I have cited. Those who review my citations will quickly realize that I stand in the Reformed Christian tradition, especially informed by people working in the tradition of Abraham Kuyper. This tradition, sometimes referred to as neo-Calvinism, has produced fruitful contributions by looking at the world through the biblical themes of creation, fall, redemption and restoration. In fact, these exact themes define the central chapters in this book.

There are many participants in the making of any book, and this is no exception. I am extremely thankful to many colleagues at Redeemer University College who encouraged, mentored and shared their time and thoughts. The exercise of writing this book was helpful in the ongoing de-

---

[1]Donald E. Knuth, *Things a Computer Scientist Rarely Talks About* (Stanford, CA: Center for the Study of Language and Information, 2001), p. 2.

velopment of my own thoughts; consequently, I hope that it will also be helpful to others who seek to understand what it means to be a faithful presence in a technological society.

I am thankful to the late Theo Plantinga for many informal discussions and encouragement to write, even though I was still in my "literary underwear" with respect to writing on this topic. Thanks to colleagues at Redeemer University College such as Wytse van Dijk, Kevin Vander Meulen, Henry Brouwer, David Koyzis, Dirk Windhorst, Al Wolters, Harry Van Dyke, Gene Haas and Syd Hielema, all of whom provided helpful input and valuable feedback. Thanks also to Peter van Beek, who provided helpful comments and feedback on the manuscript. I am grateful to Angela Bick and Marie Stevens, who patiently read through my manuscript and provided valuable editing help and stylistic suggestions. I am thankful for comments from some of my students, who were exposed to early drafts of this book. I am grateful for computer science professors at other Christian colleges who showed interest and support for this project. I am thankful to Redeemer University College for providing me with many opportunities to develop as a Christian scholar and for their support provided in many ways for this project.

I am thankful to the staff of InterVarsity Press who helped make this book a reality. In particular, I am grateful to editors Gary Deddo and David Congdon for their helpful and encouraging correspondence throughout the process. I am also grateful to several anonymous readers who were approached by the publisher and who provided many helpful suggestions to improve this book.

I am also thankful to my family, and in particular, to my wife, Carina, for her love, encouragement and support. In addition, she provided numerous helpful and practical editing suggestions, for which I am grateful.

But most of all, thanks be to God, who made all things and who continues to care for his people and his world and who will, one day, make all things new.

1

Introduction

What does Athens have to do with Jerusalem?

TERTULLIAN

■■■

Handheld communicators and doors that opened by themselves: these were some of the objects that the original producers of *Star Trek* used to portray the future. Today we have portable cell phones, automatic doors and many new developments that were not even dreamed about in early science fiction. In the Western world, we are daily dependent on a plethora of embedded computers that surround us: digital alarm clocks, computerized kitchen appliances, the myriad of processors that control our cars, our heating and ventilation systems, cell phones and, of course, personal computers. We live in a digital age in which it has become commonplace to communicate rapidly over vast networks and routinely visit websites from distant places. Computer technology has brought dramatic changes to factory floors, offices, classrooms and homes.

Does the ancient Christian faith still have anything to say to a fast-paced modern world shaped by such technology? Tertullian, a father of early Christian literature, once posed the question, "What does Athens have to do with Jerusalem?" When it comes to computer technology, we might well ask, "What does Silicon Valley have to do with Jerusalem?" In a nutshell: what do bytes have to do with Christian beliefs?

This book is dedicated to working out the question of what faith has to do with computer technology. Not only is this question of academic interest; it also has many implications for a world in which computer technology has become ubiquitous. Computer technology changes so rapidly that we often do not have time to adequately reflect on its impact. This impact goes beyond the tools we use; it changes the way we think and carries with it worldview implications. A worldview, in the words of Chuck Colson and Nancy Pearcey, is essentially "the sum total of our beliefs about the world, the 'big picture' that directs our daily decisions and actions."[1] A Christian worldview with respect to computer technology is the primary focus of this book.

The fact that computer technology has progressed rapidly is evident when one surveys the relatively short history of computing. Although the term *computer* was originally a term for the people employed to perform manual calculations, it later became the term used to describe the machines that replaced them. The first computers emerged as rudimentary mechanical computing "engines," developed by Charles Babbage and other pioneers in the mid- to late nineteenth century.[2] It was not until the mid-twentieth century that computers began to develop as large electronic machines that took up a sizable portion of a room. These early "big iron" machines were initially powered using primitive vacuum tubes, which were eventually replaced by smaller, cheaper and more efficient transistors. In subsequent years, techniques were developed to place numerous transistors on a single chip, called an integrated circuit. In 1971, the Intel 4004 became the first microprocessor on a single chip, comprising over 2,300 transistors. The age of the personal computer soon followed.[3]

The first personal computer kit, the MITS Altair 8800, was made available to hobbyists in 1975. Over the following decades, transistor counts continued to advance at an exponential pace, with current microprocessors now pushing transistor counts into the billions. An observation called

---

[1]Charles Colson and Nancy Pearcey, *How Now Shall We Live?* (Carol Stream, IL: Tyndale, 1999), p. 14.
[2]Charles Babbage (1791–1871) was a computer pioneer who is credited with designing the first general purpose mechanical computer. Babbage began work on a mechanical "difference engine" to assist in computing numerical tables, and later proposed an "analytical engine" that could be programmed using punched cards, but it was never finished. He is widely regarded as the father of computers.
[3]My first computer was a Sinclair ZX-81, a small personal computer that came with 1kB of RAM (which I later expanded to an impressive 16kB). A television was used for the monitor, and programs were loaded and saved on a cassette tape recorder.

"Moore's Law" predicted the exponential growth of the number of transistors on an integrated circuit.[4] In the words of Michael Rothschild, "Since the computer-on-a-chip was invented in 1971, the cost of computing has plunged 10 million-fold. That's like being able to buy a brand new Boeing 747 for the price of a large pizza."[5] These continual leaps over the course of a few decades have brought unprecedented change.

It is clear that computer technology has undergone many advances and that these have brought many changes. But before we explore the implications of this new technology, we will first clarify what we mean by *technology*.

WHAT IS TECHNOLOGY?

The word *technology* is derived from the Greek word *technologia*, which means "the systematic treatment of an art."[6] In the nineteenth century, the word was associated with mechanical and industrial arts. In recent times, *technology* has become more narrowly associated with electronics and computers. But technology actually encompasses a broad range of human activities. Carl Mitcham describes the objects of technology broadly as including all "humanly fabricated material artifacts whose function depends on a specific materiality as such."[7] He lists types of technology objects such as clothes, utensils, structures, apparatus, utilities, tools, machines and automata.[8] Mitcham explains that *clothes* include artifacts for covering the body and that *utensils* include "instruments of the hearth and home." *Structures* include buildings, while *utilities* refer to things likes roads and power networks. An *apparatus* is described as something used to control some physical process. *Tools* are defined as instruments that are operated manually, such as a pen or a hammer. *Machines* are tools that have an external source of power but still require human input, such as an automobile. And

---

[4]Moore's Law is not really a law but an observation by Gordon Moore given in 1965. It originally predicted that in the following years the number of transistors on integrated circuits would approximately double each year. This trend continued into the mid 1970s, after which the doubling continued every eighteen months or so.

[5]Michael Rothschild, "Beyond Repair: The Politics of the Machine Age Are Hopelessly Obsolete," *The New Democrat*, July/August 1995, p. 9.

[6]Stephen V. Monsma, ed., *Responsible Technology* (Grand Rapids: Eerdmans, 1986), p. 11.

[7]Carl Mitcham, *Thinking Through Technology: The Path Between Engineering and Philosophy* (Chicago: University of Chicago Press, 1994), p. 161.

[8]Ibid., p. 162.

finally, *automata* refers to machines that require neither human energy input nor immediate human direction. Thus, the term *technology* encompasses a broad range of objects, including ones that are not often associated with the word. Indeed, clothes and utensils are types of technology even if they are not commonly recognized as such.

But computers are unique in that they are more than an apparatus, a utility or a tool. Some computer applications fall under the category of machine, since some computer operations require human interaction. Computers figure most prominently in the category of automata, however, since they are capable of functioning without human direction once they are programmed to complete a task. For example, a computer-controlled thermostat is capable of automatically regulating temperature using a program designed for the task.

Mitcham argues that technology is not just made up of types, but that it has modes of interaction. Beyond basic physical interaction with technological objects, he identifies technological knowledge, technological activities and technological volition.[9] *Technological knowledge* includes concepts such as recipes, theories, rules and intuitive "know-how." *Technological activities* include actions like design, construction and use. Finally, *technological volition* covers knowing how to use technology and understanding its consequences. These various modes demonstrate that a thoughtful definition of technology will encompass more than just the types of physical devices we use. We will take a closer look the consequences of technology before offering a definition of computer technology.

TECHNOLOGY IS NOT NEUTRAL

The concept of technological volition recognizes that technology is shaped by human will. Nevertheless, some have suggested that technology itself is neutral; it is just a tool that can be used either for good or for evil. In the words of one author, technology is "essentially amoral, a thing apart from values, an instrument which can be used for good or ill."[10] The typical argument goes something like this: it's not the technology itself but what you

---

[9]Mitcham, *Thinking Through Technology*, p. 159.
[10]Robert Angus Buchanan, *Technology and Social Progress* (New York: Pergamon Press, 1965), p. 163.

do with technology that counts. The assumption that a technical artifact is just a neutral tool is sometimes referred to as *instrumentalism*.[11]

Although this view may appear initially self-evident, the fact is that technology is value-laden. Christian philosophers have described this notion more broadly by stating that creation not only has a structure but also a direction.[12] The designers of technological objects embed their personal or corporate values into their devices. Consequently, there is a direction embedded in the structure of technological artifacts.[13] As a result, technological objects are biased toward certain uses, which in turn bias the user in particular ways. Cultural critic Neil Postman explains the nonneutrality of technology as follows: "Embedded in every tool is an ideological bias, a predisposition to construct the world as one thing rather than another, to value one thing over another, to amplify one sense or skill or attitude more loudly than another." Postman goes on, "New technologies alter the structure of our interests: the things we think *about*. They alter the character of our symbols: the things we think *with*. And they alter the nature of community: the arena in which thoughts develop."[14]

Marshall McLuhan went even further when he declared the now-familiar phrase "the medium is the message," suggesting that the messages embedded in technology are more significant than any content they may be used to deliver. This not only applies to computer technology, but also to older technologies such as the printed word, the telegraph and the television. Each new medium brings with it a new way of thinking and looking at the world. In fact, the content of a medium often distracts us from the impact the technology has on us and the world around us. McLuhan put it this way, "the 'content' of a medium is like the juicy piece of meat carried by the burglar to distract the watchdog of the mind."[15] In an article explaining McLuhan's ideas,

---

[11]Nicholas Carr, *The Shallows* (New York: W. W. Norton, 2010), p. 46. See also John Dyer, *From the Garden to the City* (Grand Rapids: Kregel, 2011), pp. 84-85.

[12]Albert M. Wolters, *Creation Regained* (Grand Rapids: Eerdmans, 1985), p. 49.

[13]Charles Adams, "Formation or Deformation: Modern Technology and the Cultural Mandate," *Pro Rege* (June 1997): 3.

[14]Neil Postman, *Technopoly: The Surrender of Culture to Technology* (New York: Vintage Books, 1993), pp. 13, 20.

[15]Marshall McLuhan, *Understanding Media: The Extensions of Man* (New York: McGraw Hill, 1964), p. 18.

John Culkin writes, "We shape our tools and thereafter they shape us."[16]

It is easier to recognize the value-laden nature of technological artifacts such as handguns, nuclear bombs and land mines. These objects are obviously designed for certain purposes. Carl Mitcham makes the wry observation that people do not use guns as toothpicks. Mitcham anticipates the argument that perhaps nuclear bombs could be used for peaceful purposes such as digging canals, but he argues that such talk is "unrealistic and misleading" because bombs are "inherently oriented to military use."[17]

Many technological artifacts have values and directions that are less obvious. Consider, for instance, the invention of the mechanical clock. Neil Postman writes about the fascinating history of clocks, which Benedictine monks invented in the twelfth century. Their original purpose was to regulate devotional times.[18] But clocks mark, measure and quantify time in any domain, and they soon began regulating work, commerce and almost every part of life. Lewis Mumford makes the point that the "the clock is not merely a means of keeping track of the hours, but of synchronizing the actions of men." He continues, "The clock is a piece of machinery whose 'product' is seconds and minutes: by its essential nature it dissociated time from human events and helped create the belief in an independent world of mathematically measurable sequences."[19] Ironically, clocks were originally designed to improve devotional practices, but they ended up influencing almost every aspect of life. The direction and value-laden quality of technologies are not always easy to discern, but this fact does not make them any less real.

Computers are technological artifacts, but what values are embedded in computers? Canadian philosopher George Grant quotes a computer scientist who said, "The computer does not impose on us the ways it should be used."[20] Although this statement seems like common sense, Grant unpacks its hidden assumptions. Computer technology is definitely not neutral; it changes our world, and we are just starting to understand the extent of these changes. Grant observes, "It is clear that the ways that computers can

---

[16]John M. Culkin, "A Schoolman's Guide to Marshall McLuhan," *Saturday Review*, March 18, 1967, p. 70.

[17]Mitcham, *Thinking Through Technology*, p. 252.

[18]Postman, *Technopoly*, p. 14.

[19]Lewis Mumford, *Technics and Civilization* (New York: Harcourt, Brace, 1934), pp. 14, 15.

[20] Quoted in George Grant, *Technology and Justice* (Toronto: House of Anansi, 1986), p. 19.

be used for storing and transmitting information can only be ways that increase the tempo of the homogenizing processes. Abstracting facts so they can be stored as information is achieved by classification, and it is the very nature of any classifying to homogenize. Where classification rules, identities and differences can appear only in its terms."[21]

In other words, computers must convert information into a form they can store and represent. That process requires a type of classification that limits the range of possibilities for information that is stored in a computer. Grant gives the example of storing assessments of children's skills and behavior in a computer, and the homogenization that takes place when facts are abstracted so they can be stored as data. Storing data in a computer requires quantification, and one issue with quantification is that it reduces things to "what can be counted, measured, and weighed."[22] It is not simply a matter of whether a computer is used to do good or evil, such as making a computer virus versus sending an encouraging email. The computer changes the way we think and frame the world around us. Although there is a certain amount of latitude in how a personal computer may be used, it tends to emphasize speed and the abstraction and quantification of things.

Quantification and abstraction are powerful tools in engineering and computer science, but they must never be confused with the reality they represent. One must avoid *abstractionism*, which is "the belief that our theoretical abstractions from reality are true representations of reality."[23] Jaron Lanier, a computer scientist and pioneer in the field of virtual reality, observes, "Information systems need to have information in order to run, but information underrepresents reality."[24] Computer scientist Frederick Brooks writes that "models are intentional oversimplifications to help us with real-life problems that are frighteningly complicated." He warns that "the map is not the terrain" and that models do not form a complete picture.[25] This is a critical point, since computers rely on models and have become the primary tool

---

[21]Ibid., p. 23.
[22]Egbert Schuurman, *Technology and the Future: A Philosophical Challenge* (Toronto: Wedge Publishing, 1980), p. 344.
[23]Charles Adams, "Automobiles, Computers, and Assault Rifles: The Value-Ladenness of Technology and the Engineering Curriculum," *Pro Rege* (March 1991): 3.
[24]Jaron Lanier, *You Are Not a Gadget* (New York: Knopf, 2010), p. 69.
[25]Frederick P. Brooks, *The Design of Design* (Boston: Addison-Wesley, 2010), p. 33.

with which we analyze and communicate ideas. While some types of information can be easily represented in a computer, other areas are not so easily
quantified and are ill-suited to analysis by a computer.

Values are also implicit in the problems that computer programmers
choose to solve. The fact that software is designed to solve a particular
problem presupposes a certain set of beliefs regarding the problem being
solved. For example, the SETI@home project uses Internet-connected computers in the Search for Extraterrestrial Intelligence (SETI) by analyzing
radio telescope data.[26] These efforts presuppose a certain set of "control beliefs" about the possibility of extraterrestrial intelligence. Christian philosopher Nicholas Wolterstorff describes *control beliefs* as those beliefs about
reality that enable us to commit to a theory.[27] For instance, bio-engineering
software is preceded by a set of beliefs about life and the extent to which
technological manipulation is permitted. In general, the technological
projects that people or corporations pursue are often things that are important to them: things that they value and believe are worthwhile or true.

The World Wide Web is another example of a technology that is not
neutral. The web has challenged the notion of authoritative sources and
the meaning of truth.[28] More critically, the web as a medium encourages us
to "surf" rather than dive down deeply into reflective reading. In a sea of
hyperlinks, we tend to scan text and images and flutter from one link to
another. In his provocatively titled article "Is Google Making Us Stupid?"
Nicholas Carr laments, "Once I was a scuba diver in the sea of words. Now
I zip along the surface like a guy on a Jet Ski."[29] Rapid access to vast amounts
information and the speed of information interchange has increased the
pace of business and life. Some neuroscientists are even suggesting that the
medium of the Internet is altering the way young brains are developing
and functioning.[30] The medium of the web has done more than just deliver

---

[26]For more information, see http://setiathome.berkeley.edu.

[27]Nicholas Wolterstorff, *Reason Within the Bounds of Religion* (Grand Rapids: Eerdmans, 1999), pp. 67-68.

[28]For example, see Simson L. Garfinkle, "Wikipedia and the Meaning of Truth: Why the On-line Encyclopedia's Epistemology Should Worry Those Who Care About Traditional No-tions of Accuracy," *MIT Technology Review* (November/December 2008).

[29]Nicholas Carr, "Is Google Making Us Stupid?" *The Atlantic*, July/August 2008, p. 57.

[30]Gary Small, *iBrain: Surviving the Technological Alteration of the Modern Mind* (New York: William Morrow, 2008).

information in a new way; it appears to be changing the very way we think. Expanding on his article, Carr later wrote a book titled *The Shallows,* which explores these issues in greater depth.[31] In his book he states, "Dozens of studies by psychologists, neurobiologists, educators, and Web designers point to the same conclusion: when we go online, we enter an environment that promotes cursory reading, hurried and distracted thinking, and superficial learning."[32] Author Tim Challies warns that constant distraction leads to shallow thinking, and that shallow thinking leads to shallow living.[33] When evaluating computer technology, we should not only ask what new things it makes possible, but also *what is made more difficult or perhaps even impossible.*[34]

Computer-driven areas such as virtual reality (VR) and robotics are media that have a message as well. Virtual reality and computer games simulate the experience of the real world and allow users to create their own worlds. With the addition of multimodal devices such as motion gloves, head-mounted displays and even tactile feedback using haptic devices, the experience of virtual reality is amplified as more senses are included.[35] The medium of VR will have profound changes on the way people view and experience reality itself.[36] As virtual reality becomes more compelling, the notion of what is actually real will also begin to change. Likewise, humanoid robotics and cyborgs have begun to raise questions about the differences between humans and machines. In her book *Alone Together* Sherry Turkle argues that "thinking about robots . . . is a way of thinking about the essence of personhood."[37] These advancements related to computer technology are not neutral; they embed messages that push us to see the world, and ourselves, in new ways.

---

[31]In his book, Nicholas Carr includes a section in which he shares his own challenges dealing with digital distractions while trying to concentrate on writing the book.

[32]Nicholas Carr, *The Shallows* (New York: W. W. Norton, 2010), pp. 115-16.

[33]Tim Challies, *The Next Story: Life and Faith After the Digital Explosion* (Grand Rapids: Zondervan, 2011), p. 117.

[34]Andy Crouch, *Culture Making* (Downers Grove, IL: InterVarsity Press, 2008), pp. 29-30.

[35]Haptic technology provides tactile feedback to a user so mechanical forces can be sensed to make the control of virtual objects more realistic or to assist in the remote control of robots and devices (telerobotics).

[36]Schuurman, *Faith and Hope,* 129.

[37]Sherry Turkle, *Alone Together: Why We Expect More from Technology and Less from Each Other* (New York: Basic Books, 2011), p. xvii.

Marshall McLuhan identified four "laws of media" that summarize how media and artifacts, including technological artifacts, exert an influence on us. These four laws can be posed as questions as follows:[38]

1. What does the artifact *extend* or *enhance*? What human capacity is amplified?

2. What does the artifact make *obsolete*?

3. What does the artifact *retrieve* from the past?

4. When pushed to its limits, an artifact tends to *reverse* its original characteristics. What does the artifact reverse into?

To illustrate how these four questions are used, McLuhan applies them to a variety of different media and technologies. For example, McLuhan applies them to the technology of the car.[39] In answer to the first question, the car *enhances* privacy and mobility. The second question considers what an artifact makes *obsolete*; in the case of the car, it is the horse and buggy. The third question considers what an artifact *retrieves* from the past. McLuhan suggests that the car retrieves from the past the notion of a "knight in shining armor." The fourth question considers some of the unintended consequences when an artifact is pushed to the limit. In the case of the car, when it is used heavily, it leads to a *reversal* of mobility—namely traffic jams and congestion. These four questions can be applied to different technologies and can be helpful in identifying some of the effects of a technology and uncovering ways in which it is not neutral.

TECHNOLOGY AND TECHNIQUE

In the mid-twentieth century, French philosopher and sociologist Jacques Ellul described the wide-ranging impact of technology by using the term *technique*. In *The Technological Society*, Ellul defines technique as "the totality of methods rationally arrived at, and having absolute efficiency (for a given stage of development) in every field of human activity."[40] For Ellul, technique is a mindset in which all things are problems to be solved using

---

[38]Marshall McLuhan and Eric McLuhan, *Laws of Media: The New Science* (Toronto: University of Toronto Press, 1988), pp. 98-99.

[39]Ibid., p. 148.

[40]Jacques Ellul, *The Technological Society* (New York: Vintage Books, 1964), p. xxv.

efficient methods. The worldview of technique, with its focus on efficiency, has been applied to every field of human activity including death, procreation, birth and habitat.[41] This mindset is so pervasive that it has even been applied to the church. The attitude of technique is evident in the abundance of how-to books that promise proven methods for growing the church, developing leaders and expanding ministries—as if such things could be reduced to a formula. Technique is about more than technological artifacts; it is about a way of thinking.

Furthermore, in *The Technological Society*, Ellul suggests that choice disappears as technology measures the best means based strictly on efficiency and "numerical calculation." "No human activity escapes this technical imperative," Ellul asserts.[42] The notion of the *technological imperative* suggests that once technological developments are underway, they are unstoppable. Ellul says, "If a desired result is stipulated, there is no choice possible between technical means and non-technical means. . . . Nothing can compete with technical means. The choice is made *a priori*. It is not in the power of the individual or the group to decide to follow some method other than the technical."[43]

A modern example is that of the automobile. Cars have provided freedom and mobility, but they have fundamentally changed our neighborhoods, our cities and the way we live. In many places, the choice to walk has become difficult and unpleasant, with infrastructure primarily constructed to accommodate people who travel in cars. Most people commute alone in their cars on vast freeways and have fewer opportunities to encounter their neighbors and experience community than previous generations did. As mentioned earlier, when cars are used heavily, a reversal occurs, resulting in traffic jams and reduced mobility due to congestion. Other problems include rising pollution and accidents. Even so, the car has become indispensable in our society.

Similarly, computer technology has become a requirement to function in our society. In the home, workplace and school, the computer has become an indispensable tool. Many retailers and services are now only available on the

---

[41]Ibid., p. 128.
[42]Ibid., p. 21.
[43]Ibid., p. 84.

web or in an electronic format. Even the choice of the software we run on our computers is often dictated by forces around us. The computer has left in its wake various problems and challenges, but its necessity is now a foregone conclusion. With each new technology, we are quick to embrace the new possibilities it brings and sometimes think little about what we might be losing.

Neil Postman introduces a similar notion with the term *technopoly*. In a book titled after this term, Postman defines *technopoly* as "the submission of all forms of cultural life to the sovereignty of technique and technology."[44] This view, sometimes referred to as *technological determinism*, sees technology as an autonomous force beyond our control.[45]

WHAT IS COMPUTER TECHNOLOGY?

Both Ellul and Postman are insightful in their analysis of the role and ubiquity of technique in modern life. But technology is not autonomous. Egbert Schuurman, a Christian philosopher of technology, responds to technological determinism by arguing that "the future of technology is in fact not determined, but open."[46] It is humans who have responsibility for how technology unfolds. The book *Responsible Technology* captures this notion well by defining technology as "a distinct cultural activity in which human beings exercise freedom and responsibility in response to God by forming and transforming the natural creation, with the aid of tools and procedures, for practical ends or purposes."[47] This definition captures a number of important points while avoiding the pitfalls of both instrumentalism and determinism. Technology is not neutral; it is a value-laden cultural activity in response to God that shapes the natural creation. Neither is technology autonomous; it is an area in which we exercise freedom and responsibility.

Since the focus of this book is computer technology, we will modify this definition to focus specifically on the computer. A *computer* can be defined as an electronic device that receives input, processes and stores data according to a program, and produces output. But computer technology is more than the study of computers; to put it differently, "Computer science

---

[44]Postman, *Technopoly*, p. 52.
[45]Carr, *The Shallows*, pp. 46-47.
[46]Schuurman, *Technology and the Future*, p. 361.
[47]Monsma, *Responsible Technology*, p. 19.

is no more about computers than astronomy is about telescopes."[48] The study of computer technology comprises not only the hardware and the physical machine that performs the processing, but also the software and the exploration of the possibilities of computation. Consequently, computer technology may be defined as: *a distinct cultural activity in which human beings exercise freedom and responsibility in response to God, to unfold the hardware and software possibilities in creation with the aid of tools and procedures for practical ends or purposes.*

This definition of computer technology is an adaptation of the previous definition of technology, which encapsulates several important elements. First, technology is a *human cultural activity*; it is more than just products and devices. Author Andy Crouch describes culture as "what human beings make of the world," and technology is part of that activity.[49] Second, it recognizes that computer technology is a *response to God*, which implies we have a responsibility. The response can be obedience to God's will or disobedience and rejection of God. The next part of the definition identifies both *hardware and software* as the two main components specific to computer technology. Computer technology includes physical realizations, such as the electrical and mechanical structure of computers, as well as more abstract and intangible aspects, such as software. The next phrase in the definition suggests that computer technology is not a naturally occurring phenomenon, but rather is constructed *with the aid of tools and procedures*. With computer technology, the tools include software tools like compilers and editors as well as hardware tools like logic analyzers and soldering irons. The reference to procedures indicates that there are certain processes and expert knowledge required in the development of computer technology. For instance, programming requires an *algorithm*, a step-by-step method for solving a particular problem; it is the computer equivalent of a "recipe" for solving a problem. Furthermore, the manufacture of computer chips requires complex tools and procedures to transform wafers of silicon into functioning digital circuits.

Finally, this definition ends with the statement that computer technology

---

[48]Michael R. Fellows and Ian Parberry, "SIGACT Trying to Get Children Excited About CS," *Computing Research News* 5, no. 1 (January 1993): 7.

[49]Crouch, *Culture Making*, p. 23.

is done for *practical ends or purposes*. Frederick Brooks emphasizes the practical nature of the discipline by stating that computer science is a synthetic discipline "concerned with making *things*, be they computers, algorithms, or software systems."[50] As such, technology is fundamentally different from aesthetic or contemplative activities. As a synthetic discipline, it is also different than a pure science. Brooks summarizes it this way: "the scientist *builds in order to study*; the engineer *studies in order to build*."[51]

APPROACHES TO COMPUTER TECHNOLOGY

The definition we have established for computer technology states that it is a distinct cultural activity. There has been much discussion on the topic of Christianity and its relationship to culture. In *Christ and Culture*, Richard Niebuhr summarizes several possible approaches to culture that Christians have expressed throughout history.[52] The different approaches Christians take to technology mirror these historical approaches to culture. The possible responses that Christians take to technology include the following:

- rejection of technology

- indifference to technology

- embracing technology

- cultivating responsible technology

You may be able to think of people you know who exemplify each of these categories. We will briefly look at the first three approaches before turning to the fourth approach, which is the one advocated in this book.

Rejection of technology. Those who reject technology or view it with disdain are sometimes labeled *technophobes* or *neo-Luddites*. The Luddites originated in the early nineteenth century, when a group of disgruntled textile workers burned and destroyed factories in protest against the perceived threat of mechanization. The term *Luddite* came from General Ned Ludd, who was the fictitious leader of their movement.[53] Throughout history, there have been

---

[50]Frederick P. Brooks, "The Computer Scientist as Toolsmith II," *Communications of the ACM* 39, no. 3 (March 1996): 62.

[51]Ibid.

[52]H. Richard Niebuhr, *Christ and Culture* (New York: Harper & Row, 1951).

[53]Sara Baase, *A Gift of Fire: Social, Legal, and Ethical Issues for Computing Technology*, 4th ed. (Upper Saddle River, NJ: Prentice Hall, 2013), p. 334.

people who have rejected technology, perceiving its effects as undesirable or perhaps even a threat. In *Walden*, which recounts his experiences living in a cabin in the woods for two years, Henry David Thoreau remarks that our inventions are but "improved means to an unimproved end."[54] To this day, the Amish and Old Order Mennonites choose to live simple lives, pursuing traditional, rural lifestyles without modern technology. In his book *Better Off*, author Eric Brende describes a real-life experiment in which he and his wife move to a remote, rural Amish community for eighteen months. The book is an engaging account of living without the aid of modern technology. Brende's conclusion is that although technology makes things easier, life might be preferable with less technology.[55]

Such views are out of step with the prevailing twenty-first-century culture, but they are not necessarily ill-informed or crazy. For instance, Wendell Berry has written a provocative essay titled "Why I Am Not Going to Buy a Computer."[56] In this essay Berry, a prolific writer, describes his preferred method of farming as one that uses horses, and his preferred writing process as one that uses a paper and pencil (with the help of his wife, who types his manuscripts). Berry resists the impulse to throw out the "old model" and provides a list of guidelines for adopting new technologies. He claims that computers do not improve his writing and that, in fact, they have several disadvantages. These disadvantages include expense, size and electricity requirements (which are not necessarily generated from clean power sources). Although Berry is a thoughtful writer, these arguments would not convince most people. In the twenty-first century, neo-Luddites are increasingly rare.

Indifference to technology. The second approach is one in which people are simply indifferent to technology. According to this approach, bytes don't have much to do with Christian beliefs. Richard Niebuhr describes people who are indifferent to culture as "[feeling] equally at home in the community of culture. They feel no great tension between church and world, the social laws and the Gospel, the workings of divine grace and human effort."

---

[54]Henry David Thoreau, *Walden* (Princeton, NJ: Princeton University Press, 1973), p. 52.
[55]Eric Brende, *Better Off: Flipping the Switch on Technology* (New York: HarperCollins, 2004), p. 229.
[56]Wendell Berry, *What Are People For? Essays by Wendell Berry* (Berkeley, CA: Counterpoint, 1990), p. 170.

In essence, such a view seeks to "harmonize Christ and culture."[57] This view, however, fails to recognize that a spiritual battle is raging between good and evil (see Ephesians 6:12). Technology is not neutral, and it is so pervasive and its effects are so widespread that it would be foolish not to develop a thoughtful approach to technology and culture.

Embrace of technology. The third approach is to simply embrace computer technology without much thought or reflection. This is easy to do in a society driven by the technological imperative, in which new advances in computer technology are quickly adopted without question. In this approach people become mere consumers of technology, a posture that is common with many other forms of culture.[58] Some people are more than mere enthusiasts; they see technology as the potential savior of the human condition. Egbert Schuurman notes that people who trust in technology "never notice their own slavery" because it is "obscured by the anesthetizing influence of technology's possibilities."[59] Like a fish in water, people are not always aware of the extent to which the prevailing technological society has shaped the way they live and think.

Cultivating responsible technology. A thoughtful approach to computer technology needs to be more nuanced than just rejecting it out of hand or simply embracing every new development that comes along. Simplistic views of technology either overestimate its ability to solve human problems or blame it for all our problems. Both these views portray technology as an independent force rather than emphasizing the fact that it is a human activity. If technology is, in fact, *a distinct cultural activity in which human beings exercise freedom and responsibility in response to God,* then we need to use and develop computer technology in ways that honor God. This is the essence of the fourth approach: responsibly engaging computer technology. If technology is an extension of creation, we must be able to use it in a way that glorifies God and furthers his kingdom.[60] We cannot do that unless we pay attention to the "direction" of technology. This is something that individuals, together with the wider Christian community, need to

[57]Niebuhr, *Christ and Culture,* p. 83.
[58]Crouch, *Culture Making,* pp. 89-90.
[59]Egbert Schuurman, *Technology and the Future,* p. 368.
[60]Theodore Plantinga, *Rationale for a Christian College* (Grand Rapids: Paideia Press, 1980), p. 57.

discern. The quest for a responsible Christian approach to computer technology begins by looking at the Bible, which provides a light for our path (see Psalm 119:105).

But how do we use the Scripture to light our way when we are traveling along new paths? A good place to start is to consider the main biblical themes of creation, fall, redemption and restoration. Each of these themes has implications for how we approach computer technology. In creation we recognize God as the Creator of the heavens and earth, including the latent potential for computing. Tragically, the human family brought sin into the world, and this has implications for all of creation, including human cultural activities like computer technology. Thankfully, God did not abandon us to despair but sent Jesus Christ to redeem his people and his world. Jesus has inaugurated his kingdom on earth, and one day he will return to make it altogether good again. Until then, God's children are to, in the words of Lewis Smedes, "go into the world and create some imperfect models of the good world to come."[61] The remainder of this book will examine in detail how a worldview informed by these biblical themes can help guide Christians who seek to honor God in the area of computer technology.

---

[61]Lewis Smedes, *My God and I* (Grand Rapids: Eerdmans, 2003), p. 59.

2

Computer Technology and
the Unfolding of Creation

Why is programming fun? What delights may its practitioner expect
as his reward? First is the sheer joy of making things.

FREDERICK BROOKS, *THE MYTHICAL MAN-MONTH*

■■■

Computer vision is a fascinating area of study that deals with extracting meaningful information from digital images. While the human vision system does this well and seemingly effortlessly, designing a reliable computer vision system remains a difficult challenge. Extracting meaning from a digital image essentially boils down to interpreting an array of pixels that represent the brightness of points in an image. In computer vision, the "low-level" or "early vision" features such as edges and line segments are relatively easy to identify. The bigger problems in computer vision involve higher-level operations such as object recognition and image understanding. The task of image interpretation is difficult, in part, because it requires insight and contextual understanding. It requires the ability to take visual cues and associate them with other knowledge.

Like computer vision, a perspective of computer technology requires a context and interpretation. The technical details—the low-level 1's and 0's, if you will—are observable by all practitioners in the field. People have different viewpoints, however, on the meaning and purpose behind the details they observe based on various presuppositions. Many people in the Western world deny the reality of a Creator and insist that the only authentic knowledge is scientific knowledge. Others acknowledge glimpses of the

divine in science, like Thomas Edison, who declared, "When you see everything that happens in the world of science and in the working of the universe, you cannot deny that there is a 'captain on the bridge.'"[1] As Christians,
we confess that the God of the Bible is, so to speak, the captain on the bridge
who created the heavens and the earth. Christians observe these same
workings of the universe and discern the fingerprints of a providential God.
Christian perspectives, including a perspective on computer technology, are
shaped by a biblical worldview.

The Heavens Declare the Glory of God

Psalm 19 opens with the words, "The heavens declare the glory of God; the
skies proclaim the work of his hands" (Psalm 19:1). This theme is expressed
well in Article 2 of the Belgic Confession, which states, "The universe is
before our eyes like a beautiful book in which all creatures, great and small,
are as letters to make us ponder the invisible things of God."[2] Similarly,
sixteenth-century Reformed theologian John Calvin wrote that God "daily
discloses himself in the whole workmanship of the universe. As a consequence, men cannot open their eyes without being compelled to see him."
Calvin continues, "There are innumerable evidences both in heaven and on
earth that declare his wonderful wisdom; not only those more recondite
matters for the closer observation of which astronomy, medicine, and all
natural science are intended, but also those which thrust themselves upon
the sight of even the most untutored and ignorant persons, so that they
cannot open their eyes without being compelled to witness them."[3] Calvin
speaks of God's glory evident in "recondite matters" such as astronomy and
the natural sciences; were he alive today, he might observe how computer
technology also demonstrates God's creative wonders.

God reveals himself in different ways, including through the Bible, which
records God's communication to us in words (his *special revelation*) and

---

[1] Quoted in David F. Noble, *The Religion of Technology: The Divinity of Man and the Spirit of Invention* (New York: Penguin, 1999), p. 97.

[2] *Ecumenical Creeds and Confessions* (Grand Rapids: CRC Publications, 1988), p. 79. The Belgic Confession was authored in the sixteenth century, arising out of the Protestant Reformation. It has since been adopted in many Reformed churches.

[3] John Calvin, *Institutes of the Christian Religion*, vol. 1, ed. John T. McNeill, trans. Ford Lewis Battles (Philadelphia: Westminster, 1960) (Bk 1, 5.1-2), pp. 52-53.

through the works of his hands in creation (his *general revelation*). Although it is through the Bible that we learn about the person and work of Jesus Christ, God's power is readily visible in creation. Paul writes, "For since the creation of the world God's invisible qualities—his eternal power and divine nature—have been clearly seen, being understood from what has been made, so that people are without excuse" (Romans 1:20). Skies, seas, stars, mountains, trees, birds and people all speak of a wise Creator.

But creation is much more comprehensive than seas, stars, mountains, trees and birds. Creation is everything God has ordained to exist, including families, governments, justice, art and also computers. God placed within the world the latent potential for technology and computers. This includes the possibility to etch millions of transistors onto a small silicon chip and the ability for electrical signals to propagate down wires at nearly the speed of light. There is the potential to store large amounts of data on small magnetic plates and the ability to arrange numerous light-emitting devices in rows and columns to fashion visual displays. The possibilities ordained by God are not just limited to physical devices, but also the new vistas unlocked by complex computer software. These include such delightful things as computer graphics, imaginative virtual worlds, animations and games.

Many wonders are only visible to those who venture further into God's creation. In the Psalms we read that those who "went out on the sea in ships . . . saw the works of the Lord, his wonderful deeds in the deep" (Psalm 107:23-24). Today those who peer from behind telescopes or microscopes or delve into detailed study of various aspects of God's creation can also observe his wonderful deeds. In a similar manner, computer scientists are rewarded with glimpses into some of the awesome creational structures that appear at every new turn. Researchers who explore new vistas in the world of computer technology will see that, here too, the works of the Lord are wonderful. These include the aesthetic beauty of elegant and powerful algorithms, the wonder of patterns in data and the contemplation of the notion of infinity (or just the notion of really large numbers).[4] Using complex computer simulations and numerical methods, some researchers are able to vi-

---

[4]For an example, refer to Donald Knuth's discussion of *Super K* in his lectures "God and Computer Science" found in Donald E. Knuth, *Things a Computer Scientist Rarely Talks About* (Stanford, CA: Center for the Study of Language and Information, 2001), p. 171.

sualize events and processes that would be otherwise unobservable to us. The study of computer science, like other scientific pursuits, gives us a glimpse of the majesty of a powerful and wise Creator.

IMPLICATIONS OF THE CREATION STORY FOR COMPUTER TECHNOLOGY

The creation story in Genesis establishes that God is the Creator of all things, including humankind. It is a story that also tells us something about who we are, including our place and our role in creation. The following sections explore in more detail some of the implications of the creation story for the area of computing. The creation concepts that will be highlighted include the *cultural mandate*, the notion of being created in the *image of God* and the importance of *sabbath*.

The cultural mandate. In the creation story, we are introduced to what has been called our cultural mandate. This is the divine injunction that was given to humankind in Genesis 1: "God blessed them and said to them, 'Be fruitful and increase in number; fill the earth and subdue it. Rule over the fish in the sea and the birds in the sky and over every living creature that moves on the ground'" (Genesis 1:28).

What does God want us to "fill the earth" with? This phrase does not just mean that we should have numerous children (although God does tell humankind to be fruitful and multiply). We are to "fill" the earth with the products of human culture, including books, art, music, tools and—more recently—computer technology.[5] In the previous chapter, our definition of technology included the notion that it is "a distinct cultural activity in which human beings exercise freedom and responsibility in response to God."[6] At the time of creation, God made a world pregnant with possibilities and gave human beings the delightful task of opening up the potential of God's creation. In Psalm 8, we read of the great God who made the stars and the heavens and who made humankind the rulers of the works of his hands. This mandate has never been rescinded; even after the fall into sin, this task remains. Since computing is one of the possibilities in creation, we must steward this as we would any other aspect of God's creation.

---

[5]Richard J. Mouw, *Calvinism in the Las Vegas Airport* (Grand Rapids: Zondervan, 2004), p. 79.
[6]Stephen V. Monsma, ed., *Responsible Technology* (Grand Rapids: Eerdmans, 1986), p. 19.

Already in Genesis 4 we read of people beginning to uncover some of the possibilities in creation. This included people like Cain, who built the first city. Others, such as Jabal, became the father of those who raise livestock. Jubal was the father of all who play the harp and flute. There was also Tubal-Cain, who forged all kinds of tools out of bronze and iron. This was perhaps the first report of an early activity that might be called engineering. Although the possibilities for computers have been present since the time of creation, it is only relatively recently that this area has been uncovered and developed. Adam probably had no inkling of the potential that lay in the earth's materials, waiting to be discovered. This included the materials that might have been found right beneath his feet, like sand containing the element silicon, which would one day form the substrate for computer chips.

The cultural mandate includes the notion of an opening-up process. For instance, there is historical development in legal concepts, government institutions and education. Likewise, this unfolding process is evident in the history of computer technology. What started out as a branch of mathematics and electrical engineering has blossomed into its own discipline, and it has subsequently spawned newer subdisciplines that did not exist a few decades ago. These new disciplines include computer science, computer engineering and software engineering. There has also been an emergence of related disciplines focusing on the applications of this technology, such as information technology (IT) and information systems (IS). The field of computer science has further opened up into many subareas of research. These include areas such as operating systems, networks, security, artificial intelligence, databases, programming languages, compilers and computer graphics, to name just a few. It is exciting to imagine what other aspects of creation or fields of research are yet to be uncovered!

Computer technology has helped uncover new vistas and hidden treasures in other areas of creation. For example, computers have been crucial in the study of chaos and fractals. The infinite complexity and beauty of fractal patterns only became apparent when the computational and visualization capabilities of computers became widely available in the early 1980s. Computers have also helped to open new subfields in other disciplines, such as bioinformatics in biology and digital imaging in art. Likewise, recent advances in supercomputing are spurring new drug discoveries and improving our understanding of the stars and galaxies. Supercomputers also enable detailed simulations of complex processes such as the birth of stars, turbulence, earthquakes and climate.

Genesis pictures God as the King who speaks everything into existence and names what he creates. The act of naming had special significance to the Hebrew people and indicated a sovereign right.[7] As a steward of God's creation, Adam is given the job of naming the animals in the Garden of Eden. This task continues to this day, as we uncover new areas of creation and give names to various things uncovered in different disciplines of study. By naming things, we identify them, which enables us to study them further. In the area of computer technology, people have identified and named a wide variety of structures and ideas that have been discovered. In computer science, names have been given to various algorithms, like the Bubble sort and the Quicksort algorithm. Descriptive names have also been assigned to data structures—for example, trees, queues and stacks. Programmers also give names to their new software creations, sometimes showing a high degree of playfulness in their naming conventions.[8] Different reusable software solutions to commonly occurring problems have been identified as *design patterns*, and each pattern is given a name to identify it.[9] Some computing-related terms have arisen from their acronyms, such as RAM (Random Access Memory), ROM (Read-Only Memory), ALU (Arithmetic Logic Unit) and CPU (Central Processing Unit).[10] The proliferation of new names and acronyms in computer technology sometimes reduces to jargon, which leads to confusion rather than understanding. But in general, the act of naming something goes hand in hand with being stewards unfolding God's creation.

The ability to use names in computer science also enables abstraction. Certain pieces of program functionality can be gathered together and iden-

---

[7]Craig G. Bartholomew and Michael W. Goheen, *The Drama of Scripture* (Grand Rapids: Baker Academic, 2004), p. 34.

[8]Many examples of this are familiar to computing enthusiasts, such as the recursive acronym *GNU* (which stands for *GNU's Not Unix*).

[9]To learn more about design patterns, refer to Erich Gamma et al., *Design Patterns: Elements of Reusable Object-Oriented Software* (Boston: Addison-Wesley, 1994).

[10]There is certainly an abundance of TLAs (Three-Letter Acronyms) in computer technology!

tified by a name. Whenever this piece of functionality is required, it can simply be referred to by name rather than repeating all the details. Through the use of names, the underlying implementation details can be hidden, thereby reducing complexity. Some programming languages make extensive use of software entities called *objects*, in which a name (or *identifier*) can be used to represent a software bundle of related data and instructions.[11] These objects are used to perform various tasks while their underlying implementation details can remain hidden. As more complex programs are built up, multiple layers of abstraction can be employed in the design and implementation. Thus names are not only used to classify new concepts and discoveries, but to help build more complex structures and programs. The activity of naming in computing, just as Adam named the animals, is one activity associated with the cultural mandate.

Sadly, the cultural mandate has at times been misused to justify exploiting and plundering the world and its resources. In fact, it has been argued that the blame for Western attitudes toward nature ought to be laid at the feet of Christianity. An influential article by Lynn White, titled "The Historical Roots of Our Ecological Crisis," blames Christianity for our current environmental crisis.[12] White claims that the creation story discarded previous notions about the sacredness of nature and led to attitudes of mastery and dominion through science and technology. His analysis is a sobering reminder that Christians have not always been at the forefront of stewardship and earthkeeping. In many ways, we have "tilled the earth," but we often forget the second half of the cultural mandate: to also take care of it.[13] Proper earthkeeping, done according to God's normative standards, will result in shalom for the creation. It is because of sin and selfishness, to which Christians are not immune, that the earth has not been cared for in a way that reflects God's intent. Computer technology has further strained the environment, with increases in power consumption and improper disposal and poor recycling of electronic materials. The constant drive to continually upgrade and update computers and cell phones has left us with the

---

[11]This type of programming is often referred to as Object-Oriented Programming, sometimes using the acronym OOP.

[12]Lynn White Jr., "The Historical Roots of Our Ecological Crisis," *Science* 155, no. 3767 (March 1967): 1203-7.

[13]Bob Goudzwaard, *Idols of Our Time* (Downers Grove, IL: InterVarsity Press, 1984), p. 107.

ongoing problem of how to deal with the castoffs of yesterday's technology.

When God spoke to the Israelites about the Promised Land, he not only spoke of a land flowing with milk and honey, but a "land where the rocks are iron and you can dig copper out of the hills" (Deuteronomy 8:9). Iron and copper and other metals are necessary for technological progress. God promises them a land where food is not only plentiful, but where they can mine and develop tools. This promise, however, is followed by a warning not to forget the Lord nor to attribute personal wealth to their own power and strength (see Deuteronomy 8:14, 17). Instead, God's people are called to remember the God "who gives you the ability to produce wealth" (Deuteronomy 8:18). God also equips people with the ability to understand creation; in Isaiah, God even teaches the farmer how to till the soil (see Isaiah 28:23-29). As God's stewards, we need to seek God's help to discern how to use the earth's resources and develop tools to serve him responsibly.

The image of God. The creation story clearly states that humans were created in the "image of God," whereas birds, fish and animals were created "according to their kinds" (Genesis 1:27, 21, 24). This would also distinguish humans from machines. But what does it mean to be made in the image of God? Is it primarily our ability to think or reason that sets us apart? For John Calvin, the image of God was essentially spiritual in nature and "extends to the whole excellence by which man's nature towers over all the kinds of living creatures."[14] It is difficult to articulate a precise interpretation of what is meant by the *image of God*. However, it embeds some basic concepts about how humans are distinct from other parts of the creation, including machines and computers. Ultimately, we need to look to Jesus Christ, who is the true image of God (see Colossians 1:15). By becoming more Christlike, we learn what it means to be faithful image-bearers of God.

Being made in the image of God has implications for how we develop and use computer technology. As image-bearers, we have been given responsibility over creation and we are to live in loving communion with each other (see Genesis 1:28). We ought to use computer technology to show love to our neighbor and in service of all kinds of life. Because we are made to live in community with others, it is not healthy, for example, to allow elec-

---

[14]Calvin, *Institutes*, vol. 1 (Bk 1, 15.3), p. 188.

tronic communications to replace most of our face-to-face interactions. What is more, we must not seek to use technology to become like God—we are only images of God, and fallen images at that. In addition, because we are made in God's image, we ought to reject a materialistic reduction of what it means to be human.[15] We are more than machines.

Being made in the image of God also implies that we have the capacity to be creative. The notable computer scientist Frederick Brooks expresses this notion well in his classic book, *The Mythical Man-Month*: "Why is programming fun? What delights may its practitioner expect as his reward? First is the sheer joy of making things. As the child delights in his mud pie, so the adult enjoys building things, especially things of his own design. I think this delight must be an image of God's delight in making things, a delight shown in the distinctness and newness of each leaf and each snowflake."[16]

The creator of the Linux operating system, Linus Torvalds, in his autobiographical book titled *Just for Fun*, expresses a similar delight: "With computers and programming you can build new worlds and sometimes the patterns are truly beautiful."[17] Brooks observes that "the programmer, like the poet, works only slightly removed from pure thought-stuff. He builds his castles in the air, from air, creating by exertion of the imagination."[18] Indeed, anyone who has seriously programmed has likely experienced similar sentiments when they find elegant solutions to complex problems—often after a period of long and difficult work. As image-bearers of God, we can delight in creative activities such as programming "castles in the air."

The creation story also tells us something about how human beings were created male and female, and that both are made in God's image. In areas like mathematics, computer science and information technology, women remain a minority. These numbers are also reflected in the profession of engineering, where men vastly outnumber women. This is despite the fact that a woman named Ada Lovelace is widely regarded as the

[15]Cornelius Plantinga, *Engaging God's World: A Christian Vision of Faith, Learning, and Living* (Grand Rapids: Eerdmans, 2002), p. 41.

[16]Frederick P. Brooks, *The Mythical Man-Month* (San Francisco: Wiley, 1995), p. 7.

[17]Linus Torvalds and David Diamond, *Just for Fun: The Story of an Accidental Revolutionary* (New York: HarperCollins, 2001), p. 75.

[18]Brooks, *The Mythical Man-Month*, p. 7.

world's first computer programmer. In *Unlocking the Clubhouse: Women in Computing*, the authors suggest that the male-dominated area of computing has fostered pedagogical practices and cultures that discourage talented women from pursuing studies in this area.[19] They also observe that women generally take a different approach to computing and are more interested in linking computer science to social concerns and caring for people. These observations may simply be a reflection of some of the creational differences between men and women.[20] In short, both men and women are created in God's image, and the increased participation of women in computer science can help to enhance and expand the field of computer technology.

Sabbath rest. Another aspect of the creation story is to recognize that God rested on the seventh day. In Genesis we read that "on the seventh day he rested from all his work. Then God blessed the seventh day and made it holy, because on it he rested from all the work of creating that he had done" (Genesis 2:2-3). This established the pattern of a sabbath day, which is one day in seven set aside for rest. In fact, keeping sabbath is explicitly included as one of the Ten Commandments. God's people are commanded to "observe the Sabbath day by keeping it holy" and that "on it you shall not do any work" (Deuteronomy 5:12, 14). The commandment to cease work explicitly includes children, servants, strangers and even animals. The two passages in the Scriptures that recount the Ten Commandments actually provide two different reasons to keep the sabbath. In Exodus 20, the sabbath is associated with God resting on the seventh day in creation. In Deuteronomy 5, we are told that the sabbath is to remember that God rescued his people from slavery. The reasons cited in these passages are positive ones associated with rest and reflection.

Taking time for rest and reflection has become more difficult in the computer age. Computer technology is very good at providing continual streams of information and nearly instantaneous communications. For this reason,

---

[19]Jane Margolis and Allan Fisher, *Unlocking the Clubhouse: Women in Computing* (Boston: MIT Press, 2001).

[20]Kim P. Kihlstrom, "Men Are from the Server Side, Women Are from the Client Side: A Biblical Perspective On Men, Women, and Computer Science," in *Proceedings of the Conference of the Association of Christians in the Mathematical Sciences* (Wheaton, IL: ACMS at Wheaton College, 2003), pp. 126-37.

it can be difficult to set electronic communications aside and unplug ourselves. Our electronic devices keep us tethered to work and entertainment and demand our constant attention. One author bemoans the fact that the digital revolution has plunged us into a state of "continuous partial attention," and in this state people "no longer have time to reflect, contemplate, or make thoughtful decisions."[21] Using the words of John Culkin, "We shape our tools and thereafter they shape us"—and our modern tools never rest.[22] As we become shaped by our always-on technology, we lose the ability to rest and reflect and enjoy sabbath.

The notion of sabbath is intensely countercultural, yet it is firmly embedded in the fabric of creation. Simply stated, ignoring the sabbath commandment is bound to have consequences. Although sabbath legalism should be avoided, thoughtful guidelines and practices can be helpful in safeguarding sabbath rest. Establishing periodic times to disconnect from the Internet and setting aside electronic devices is one way of promoting regular sabbath rest.

THE PROBLEM OF REDUCTIONISM

Philosopher Auguste Comte once said, "There is no inquiry which is not finally reducible to a question of numbers."[23] A contemporary view along these lines might assert that there is no area of study which cannot be understood using a computer running the proper algorithm. The seventeenth-century mathematician and philosopher Gottfried Leibniz believed that human reasoning could be reduced to a mathematical language and that debates could then be resolved using calculus. Leibniz wrote, "The only way to rectify our reasonings is to make them as tangible as those of the Mathematicians, so that we can find our error at a glance, and when there are disputes among persons, we can simply say: Let us calculate, without further ado, in order to see who is right."[24] The book *Geek Logik*

---

[21]Gary Small, *iBrain: Surviving the Technological Alteration of the Modern Mind* (New York: William Morrow, 2008), p. 18.

[22]John M. Culkin, "A Schoolman's Guide to Marshall McLuhan," *Saturday Review*, March 18, 1967, p. 70.

[23]Auguste Comte, *The Positive Philosophy* (New York: AMS Press, 1974), p. 58.

[24]G. W. Leibniz, *Liebniz: Selections*, ed. Philip P. Wiener (New York: Charles Scribner's Sons, 1951), p. 51.

takes an amusing look at this premise by suggesting equations that one can use to make a wide range of life decisions simply by plugging in the variables.[25] This book humorously suggests that the principles of basic algebra can take the guesswork out of areas ranging from dating and romance to careers and health.

Computer scientists and engineers who spend much of their time looking at the world through the narrow lens of logic and algorithms must avoid tunnel vision. God's world is much more complex and diverse. Even though many things possess computable attributes, they cannot simply be reduced to numbers. This notion is captured well by William Cameron, who wrote, "It would be nice if all of the data which sociologists require could be enumerated because then we could run them through IBM machines and draw charts as the economists do. However, not everything that can be counted counts, and not everything that counts can be counted."[26] The attempt to reduce created reality to something that can be computed is a form of *reductionism*.

One challenge that arises in technical schools and programs is the incredible level of specialization that emphasizes one kind of knowing above all others. This is due in part to the sheer volume of technical material that must be covered, which often leaves little or no time to reflect on broader perspectives of technology. Unfortunately, the consequence of intensely specialized programs is often that students become myopic and forget the greater context of creation. When one is immersed in the study of the computational or physical aspects of reality, it can be hard to zoom out and appreciate the breadth and diversity in creation. Teachers and professors who ignore opportunities to provide a wider perspective, however, are doing their students a disservice. Such an education does not do justice to the diversity and interconnectedness of creation. This wider perspective is easier to maintain in the context of a liberal arts and science education that reflects the notion that the created universe is composed of many aspects and many different kinds of knowing. A Christian study of computer tech-

---

[25]Garth Sundem, *Geek Logik: 50 Foolproof Equations for Everyday Life* (New York: Workman Publishing Company, 2006).

[26]William Bruce Cameron, *Informal Sociology: A Casual Introduction to Sociological Thinking* (New York: Random House, 1963), p. 13.

nology and engineering needs to avoid reductionism and respect the diversity and complexity in creation. Technical work has more than technical implications; it has substantial legal, ethical, political, social and other non-technical implications as well.[27]

A Creation with Many Aspects

In order to avoid reductionsm, we need to maintain a holistic view of created reality. One way of looking at creation is to use a helpful scheme suggested by the Dutch Christian philosopher Herman Dooyeweerd (1894–1977).[28] Dooyeweerd worked in the tradition of Abraham Kuyper, and his philosophy is a comprehensive framework for understanding the universe as something created, governed and sustained by God, who is absolutely sovereign. Furthermore, the creation is both diverse and interconnected. The diversity in creation is evident: rocks, trees, lakes, birds, fish, stars, planets and people.

As the possibilities in creation have unfolded, it has also included art, music, technology, schools, churches and governments. Dooyeweerd reflected on this rich diversity in our everyday experience and suggested that it may be examined in terms of a set of fifteen different "modalities" or "aspects of reality," as illustrated in figure 1. These modal aspects are not objects in themselves but are ways of understanding how diverse entities function in creation. These modal aspects are as follows: (1) the numeric, which is the aspect that relates to discrete quantities; (2) the spatial, which relates to geometry; (3) the kinematic, which relates to movement; (4) the physical, which relates to possession or exchange of energy; (5) the biological, which relates to living aspects of reality; (6) the psychic, which relates to feelings or emotions; (7) the analytical, which relates to making distinctions; (8) the historical, which relates to forming and developing human culture; (9) the lingual, which relates to language and the use of symbols to relay meaning; (10) the social, which relates to the interactions of human beings; (11) the economic, which relates to stewardship of economic re-

---

[27]Charles Adams, "Automobiles, Computers, and Assault Rifles: The Value-Ladenness of Technology and the Engineering Curriculum," *Pro Rege* (March 1991): 6.

[28]Herman Dooyeweerd, *Roots of Western Culture: Pagan, Secular and Christian Options* (Toronto: Wedge Publishing, 1979).

FAITH
trust and belief
ETHICAL
love and well-being
JURIDICAL
retribution, restitution
AESTHETIC
harmony and beauty
ECONOMIC
stewardship, frugality
SOCIAL
human interactions
LINGUAL
symbolic meaning
HISTORICAL
cultural development
ANALYTICAL
making distinctions
PSYCHIC
feelings or emotions
BIOLOGICAL
life, vitality
PHYSICAL
energy
KINEMATIC
movement
SPATIAL
geometry
NUMERIC
number, quantity

NORMS

LAWS

Figure 1. The Modal Aspects

sources; (12) the aesthetic, which relates to harmony and beauty; (13) the juridical, which relates to giving what is due, retribution and restitution; (14) the moral or ethical, which relates to love, care and well-being; and (15) the faith aspect, which relates to certitude, trust and belief.[29]

The exact number, labels and ordering of these aspects are not set in stone and have been a subject of ongoing discussion among some Christian philosophers. What is important for our purposes is to recognize that God created the world with many aspects and that all things hold together in Jesus Christ (see Colossians 1:17).

The order of the modal aspects is not arbitrary; rather, they follow a sequence and build on each other in order of increasing complexity. The "earlier" aspects (starting with the numeric) provide a foundation for the "later" ones (up to the faith aspect). Furthermore, the later aspects presuppose the earlier aspects and give meaning and value

---

[29]These modalities are described in Jonathan Chaplin, *Herman Dooyeweerd: Christian Philosopher of State and Civil Society* (Notre Dame, IN: University of Notre Dame Press, 2011), p. 59. What we have called the *analytical* aspect Chaplin calls the *logical* aspect; however, this can be easily confused with the concept of digital logic in computer science, so we will use the term *analytical* instead. Others have also used this label for this aspect; for example, see Andrew Basden, *Philosophical Frameworks for Understanding Information Systems* (Hershey, PA: IGI Global, 2007), p. 64.

to them. For example, the biological aspect presupposes the presence of the physical aspect, because biological entities have a physical basis.[30] The earlier aspects give the conditions of existence to those above. For instance, the biological aspect is necessary to make the social and psychic aspects possible.

We can observe the functioning of the various modalities when we work with computers. A simple example is a digital image; although it is formed using low-level binary pixel values (numeric aspect), the image that is formed enables higher aspects in a human subject. A digital image can prompt feelings of affection (psychic aspect), serve as a cultural object (historical aspect) and be used to portray symbols or text (lingual aspect). Digital images can portray friendships (social aspect) and also beauty (aesthetic aspect). An image can also evoke moral outrage or care in a human observer (ethical aspect). A digital image requires physical resources to store or print it, and it can also be sold (economic aspect). Images can also carry copyright restrictions or be used as evidence (juridical aspect). Finally, in a time when special effects and image manipulation are common, it requires trust for people to believe that what they see in a digital image is true (faith aspect). An image is formed using numeric data, but when this object is viewed by a human subject, it functions in a wide variety of other modes.

Although the modalities are interconnected, none of them may be reduced to any of the others; they are mutually irreducible.[31] The feeling modality cannot be reduced to the biological, and the biological cannot be reduced to the physical, which in turn cannot be reduced to the numeric modality. Attempting to explain everything in strictly numeric terms denies the multiplicity of creational aspects. For example, the aesthetic quality of music cannot be reduced to numeric measurements. Albert Einstein captured this well when he remarked, "It would be possible to describe everything scientifically, but it would make no sense. It would be without meaning—as if you described a Beethoven symphony as a variation of wave pressure."[32] Explaining created reality in terms of one modal aspect results

---

[30]Chaplin, *Herman Dooyeweerd,* pp. 57-58.
[31]Ibid., p. 57.
[32]Albert Einstein, *The Ultimate Quotable Einstein,* ed. Alice Calaprice (Princeton, NJ: Princeton University Press, 2010), p. 409.

in reductionism. In the case of computing, this occurs when one attempts to reduce higher aspects to the numeric aspect.

The different modalities are also distinguished by laws and norms. Whereas laws are in effect without human intervention, norms involve human freedom and responsibility.[33] The first four or five aspects, by their nature, are associated with creational laws that are fixed and must be obeyed. For example, the numeric aspect relates to mathematical laws. Likewise, the kinematic law of gravity and the biological process of photosynthesis must be followed according to the principles of each. We do not have freedom to go against the law of gravity or to change the sum of two numbers. However, the analytical aspect and later aspects have corresponding norms that involve human choices and freedom. Each of these modalities has associated with it "normative principles" or "regulative principles."[34] For instance, we have choices and freedom in the aesthetic sphere with regard to how to make art. Likewise, in the economic sphere, we may exercise choices about our economic resources. These modalities are also active when humans work with computer technology, and hence the definition we cited for technology includes the notion that it is a human activity involving freedom and responsibility. In his book describing Dooyeweerd's philosophy, Jonathan Chaplin puts it this way: "*Norms do not change, only the human responses to them.*"[35] A later chapter will explore the various normative principles relevant in the field of computer technology.

All fifteen modal aspects are involved when working with information and computer technology.[36] The relevance of the lower aspects is obvious: the numeric, spatial, kinematic and physical aspects are all part of the work of computer science and engineering. Less obvious, but just as important, are the ways the higher modal aspects are relevant to computer technology. For instance, the historical aspect cannot be ignored, since technology is developed within a specific context and cultural milieu. The linguistic aspect is an important part of computer languages, programming and doc-

---

[33]Chaplin, *Herman Dooyeweerd*, p. 62.
[34]Ibid., pp. 62-63.
[35]Ibid., p. 63.
[36]Basden, *Philosophical Frameworks*, pp. 182-83.

umentation. The social aspect is important in terms of understanding the issues related to electronic communications and social networking. The economic aspect deals with funding product development and working within the constraints of time and budgets. The aesthetic aspect comes into consideration in industrial design and graphical user interfaces. The juridical aspect deals with legal issues such as software license agreements and intellectual property. The ethical and moral aspects related to computer technology are also significant, especially when dealing with technology that is responsible for human well-being and safety. Finally, the faith aspect is important because our faith shapes our values, and our values shape our tools and the technologies that we pursue.

I have found the modal aspects to be a useful philosophical tool. The concept of the various modalities can increase awareness of the diversity of creation and help to avoid reductionism. It can also help us stay on guard against idolatry when one aspect of creation is elevated or made absolute. The notion of laws and norms is also a helpful distinction. In particular, the concept of norms serves to remind us of the many aspects in which human beings exercise freedom and responsibility. In later chapters we will take this tool out of our toolbox and use it to help explore contemporary issues in computer technology.

Creational Laws

The numeric and physical aspects figure prominently in computer technology. These aspects involve creational laws that define how things work and what is possible. For example, the laws of Boolean algebra are fundamental to the low-level operation of modern digital computers. This includes the logical operators AND, OR and NOT, as well as De Morgan's law.[37] Another example is Amdahl's law, which predicts the theoretical maximum speedup that can be realized when employing multiple processors. There are also physical theorems that apply to computing, such as Nyquist's sampling theorem, which deals with the bandwidth requirements

---

[37]De Morgan's law is used in Boolean algebra and can be represented as follows:
  NOT (P AND Q) = (NOT P) OR (NOT Q)
  NOT (P OR Q) = (NOT P) AND (NOT Q)

to reliably transmit a signal.[38] Another important field in computing that involves laws and limits is information theory, an area uncovered by Claude Shannon and others.[39]

There are also laws that establish hardware limits for computers. For instance, engineers are constrained by creational laws such as thermal laws and the speed of light. These laws set certain limits on the speed of processors and the speed at which data can be transmitted.

But does software have limits as well? In 1984, *Time* magazine quoted an editor of a software magazine who made the following statement: "Put the right kind of software into a computer, and it will do whatever you want it to. There may be limits on what you can do with the machines themselves, but there are no limits on what you can do with software."[40]

There are, in fact, limits to software. Some of these limits are intuitively obvious: for example, one cannot sort n objects in fewer than n steps, since one must look at each object at least once. Other limits and boundaries that computer scientists explore include the study of noncomputability and intractability. Intractable problems are problems that are not feasible to compute using existing computing devices. Some computer science research focuses on algorithmic strategies for dealing with problems that are intractable. One example is the classic "traveling salesman problem," which involves finding an optimal route for a salesman to visit a list of cities exactly once and return to the city of origin. This apparently simple problem turns out to be a deep and complex puzzle to solve, and it has many practical applications in diverse areas such as scheduling deliveries and circuit board manufacturing.[41] Some applications, like encryption, actually rely on the intractability of finding prime

---

[38]In essence, this theorem states that a sampled analog signal can be perfectly reconstructed from a sequence of samples if the sampling rate is at least $2B$ samples per second, where B is the highest frequency in the original signal.

[39]One key concept in information theory is the measure of entropy in a message, which is usually expressed by the average number of bits needed for storage or communication. To read more about Claude Shannon and the story of information theory, see chapter 7 of James Gleick, *The Information: A History, a Theory, a Flood* (New York: Pantheon Books, 2011), pp. 204-32.

[40]Alexander L. Taylor III, Michael Moritz and Peter Stoler, "The Wizard Inside the Machine," *Time*, April 1984, pp. 58-59.

[41]For a fascinating description of the traveling salesman problem, see William J. Cook, *In Pursuit of the Traveling Salesman* (Princeton, NJ: Princeton University Press, 2012).

factors for large numbers as a means of providing secure encryption keys. An encrypted message may theoretically be cracked by trying every possible key, but the large number of possible keys will lead to the passing of eons before the correct key is discovered.

Computer science pioneer Alan Turing developed a model of computing called a *Turing machine*. This "machine" was a theoretical device that manipulated symbols on a strip of tape according to a set of rules. The Church-Turing thesis hypothesized that everything computable is theoretically computable by a Turing machine. The Turing machine can be used to theorize about the limits of computing. One such limit that has been discovered is the famous *halting problem*.[42] Turing proved in 1936 that a general algorithm to solve the halting problem for all possible program inputs cannot exist for Turing machines. This is an example of a fundamental limit in computing known as *noncomputability*: problems for which algorithmic solutions are known not to exist.

Limits in computing also arise due to finite memory and finite numerical precision. For instance, not all real numbers can be precisely represented in a computer because the number of bits available to store digits is finite. Furthermore, the precision of numeric calculations is also finite. The finite precision of computers can lead to issues like rounding errors in computations. Part of the task of a theoretical computer scientist is to discover the laws of computation and map out the boundaries that form the limits of computation.[43]

One important point is that God himself is above his creation. Laws of computing are part of the created order and, like the other modal aspects, they do not transcend God's creation. As such, God is not subject to the laws of computing. Joel Adams, a computer science professor at Calvin College, writes, "if God is indeed infinite in power, then his power cannot be limited by the laws of computation, any more than Jesus' power to walk on water was limited by the law of gravity."[44]

---

[42]The halting problem can be stated as follows: given a program and an input to the program, determine whether the program will eventually halt when run with that input.

[43]Joel C. Adams, "Computing Technology: Created, Fallen, In Need of Redemption?" paper presented at the Conference on Christian Scholarship, For What? (Grand Rapids: Calvin College, 2001), p. 2. Available at http://cs.calvin.edu/p/christian_scholarship.

[44]Ibid., p. 3.

ARTIFICIAL INTELLIGENCE

One area that explores the limits of computing is the area of artificial intelligence (AI). AI has been described as "getting machines to do things that would be considered intelligent if done by people."[45] AI has raised profound questions about the mind and what it means to be human. These include many important philosophical and religious questions, such as:[46]

- Could computer hardware or software replicate the human brain?

- What is the connection between mind and body?

- What does it mean to be human?

- What is personhood?

- What is consciousness? Could a machine ever become self-aware?

The essence of these questions is not new. For centuries many thinkers, including seventeenth-century philosopher René Descartes, debated the nature of the human mind and sought to understand the relationship between the mind and the body (the so-called mind-body problem). The question of machines and intelligence was already contemplated by early nineteenth-century computing pioneer Ada Lovelace. She pondered the limits of a mechanical computer called the Analytical Engine.[47] In her notes she writes "The Analytical Engine has no pretensions whatever to originate anything. It can do whatever we know how to order it to perform."[48] In his famous 1950 paper titled "Computing Machinery and Intelligence," Alan Turing explored the question, "Can machines think?" In this paper he describes an "imitation game" which involves a human interrogator, another human and a digital computer. The interrogator is placed in a separate room and exchanges messages with the computer and the other human to determine which is which. Turing suggested that if the interrogator could

---

[45]Sherry Turkle, *Alone Together: Why We Expect More from Technology and Less from Each Other* (New York: Basic Books, 2011), p. 63.

[46]Giorgio Buttazzo, "Artificial Consciousness: Utopia or Real Possibility?" *IEEE Computer* 34, no. 7 (July 2001): 24-30.

[47]Ada, Countess of Lovelace (1815–1852), was born Augusta Ada Byron, daughter of the nineteenth-century poet George Gordon Byron. She took a keen interest in a mechanical computer called the Analytical Engine, proposed by inventor Charles Babbage. She made notes describing a method to perform certain calculations on the mechanical computer, which is widely regarded as the first computer program.

[48]Alan Turing, "Computing Machinery and Intelligence," *Mind* 59 (1950): 450.

not reliably identify which was the computer, "one will be able to speak of machines thinking without expecting to be contradicted."[49] In his paper, Turing presents arguments refuting several objections that might be raised about whether machines could think, including the "Lady Lovelace objection" cited above. This imitation game has since become known as the Turing Test, and Turing himself predicted that computers would pass the test by the year 2000.

In recent years, computers have triumphed against human opponents in several different spheres. For instance, in a highly publicized 1997 chess match, an IBM computer called Deep Blue defeated world chess champion Gary Kasparov.[50] Deep Blue used custom hardware to compute chess moves using a brute-force approach, analyzing hundreds of millions of chess moves per second. Later, in 2006, Russian chess grandmaster Vladimir Kramnik lost in a six-game match against a chess program called Deep Fritz, which ran on personal computer hardware. In 2011, another IBM computer named Watson competed on the quiz show *Jeopardy* and defeated the all-time biggest human winner of the show. At the time of writing, the Turing Test has yet to be reliably won by a computer, but many clever "chatbot" programs have been written to simulate intelligent conversation.[51]

An early example of a "chatbot" program was ELIZA, a computer simulation of a Rogerian psychotherapist written by Joseph Weizenbaum in the 1960s. It employed simple pattern matching techniques, but many users attributed more to the program than it merited. Weizenbaum was startled to observe that some people were emotionally drawn in to his program and that some psychiatrists even suggested that such a program might grow into an automatic form of psychotherapy.[52] Weizenbaum found these reactions to his program troubling, and later wrote a book called *Computer Power and Human Reason* to describe some of his

---

[49]Ibid., p. 442.

[50]This event is explored in a 2003 documentary film titled *Game Over: Kasparov and the Machine.*

[51]Various chatbots can be found on the web. Specialized chatbots have also been employed for commercial online help systems.

[52]Joseph Weizenbaum, *Computer Power and Human Reason: From Judgment to Calculation* (New York: W. H. Freeman, 1976), pp. 5-6.

thoughts about the limits of AI. Weizenbaum observes that while other machines have been made that mimic humans in certain ways, such as steam shovels, computers are viewed differently since they perform seemingly intelligent functions. He refutes a mechanistic view of humans, however, and argues that computers ought not to be used for tasks that require wisdom.[53]

John Searle wrote an article titled "Minds, Brains and Programs" in which he differentiates between "strong AI" and "weak AI."[54] A "strong AI" position holds that a machine, given the right program, could literally become a mind. Strong AI is based on presuppositions such as *functionalism*, which suggests that mental states are not based on the material a system is made up of, but rather what it does in response to inputs. This is the premise of the claims of the Turing Test: that comparing inputs and outputs can determine if a computer program is equal to the human brain. Such a view suggests that "computers given the right programs can be literally said to *understand* and have other cognitive states."[55] The "weak AI" position holds that machines are only able to simulate thinking and understanding and that simulation and duplication should not be confused.

Searle argues that the case for "strong AI" is mistaken and proceeds to frame his argument by introducing the "Chinese room" thought experiment.[56] In this experiment, a person who understands only English is locked in a room, and messages written in Chinese are passed into the room. Furthermore, the person has access to a comprehensive set of English instructions for manipulating strings of Chinese characters and has the ability to identify Chinese symbols by their shapes. Such a person could follow a detailed set of instructions on how to give back certain Chinese symbols with certain shapes in response to certain sorts of shapes. To those outside the room, it would appear as if someone in the room understands Chinese—even though he does not. Searle argues that a computer is essentially a symbol-processing machine; it cannot be said to understand, and

---

[53]Ibid., pp. 8, 227.
[54]John R. Searle, "Minds, Brains and Programs," *Behavioral and Brain Sciences* 3, no. 3 (1980): 417-24.
[55]Ibid., p. 417.
[56]Ibid., pp. 417-19.

therefore cannot be said to think. Computer scientist Edsger Dijkstra also rejected the question of thinking machines, suggesting the question as to whether machines can think is about as relevant as the question of whether "submarines can swim."[57]

The modal aspects described earlier can serve as a useful tool in this particular area. The higher modal aspects involve norms; hence, they require a human subject to exercise freedom and responsibility. Consequently, a computer can have no subject functions when it comes to the normative modal aspects; it can only function as an object. The modal aspects remind us that higher modalities cannot be reduced to the numeric modality, as would be required to run on a computer. Thus, a computer will never really understand the meaning of language (the lingual aspect), despite its powerful numeric and symbol-processing capabilities. Although computers can serve as thinking tools, they cannot understand or think. To suggest otherwise reveals a technicistic mindset.

The creation story teaches that human beings are distinct from the rest of creation. In Genesis we read "the LORD God formed a man from the dust of the ground and breathed into his nostrils the breath of life, and the man became a living being" (Genesis 2:7). Human life required more than just the material of the earth; it required the "breath" of God. A similar image is given in Ezekiel about dry bones being reconstituted into flesh, but the breath of the Lord is still necessary to impart life (see Ezekiel 37:1-14). Furthermore, in distinction from other creatures, humankind was created "in the image of God," whereas birds, fish and animals were created "according to their kinds" (Genesis 1:27, 24). For this reason, even if clever programming enables computers to someday reliably pass the Turing Test, the uniqueness of human beings made in the image of God should not be called into question.

Frederick Brooks has suggested that rather than striving for AI (artificial intelligence), a better approach would be IA (Intelligence Amplification). Rather than striving to build "giant brains" with AI, IA is a humbler and wiser approach of using the human mind along with a machine to solve hard problems. The idea is to build useful tools that augment human

---

[57]Edsger W. Dijkstra, "The Threats to Computing Science" (technical report, EWD898, Technical University Eindhoven, Eindhoven, Holland, 2003).

intelligence. Examples might include a pilot's assistant, a drilling adviser or a planning tool.[58]

CONCLUSION

In the beginning, God created the heavens and the earth. God created the world full of possibilities, but he left much of the unfolding of creation for humans to pursue. Elements in the creation story—such as the cultural mandate, the creation of humankind in the image of God and the concept of sabbath—all have implications for computer technology. Because computers work with the numerical aspect of creation, we need to be on guard against reductionism. Creation has many aspects that are both diverse and interconnected. We also recognize that in addition to laws, there are norms in creation that we must discern and for which we exercise responsibility. This has many implications for how we perceive, use and develop computer technology, including specific areas such as artificial intelligence. In the next chapter, we will turn to the biblical theme of the fall and explore its implications for those who study and work with computer technology.

---

[58]Frederick P. Brooks, "The Computer Scientist as Toolsmith II," *Communications of the ACM* 39, no. 3 (March 1996): 64.

3

Computer Technology and the Fall

To err is human; to really foul up requires a computer.

FARMERS' ALMANAC (1978)

■■■

In the beginning of *Technopoly*, Neil Postman recounts a legend by Plato, which was told by Socrates. In this story there is a king of a great city, called Thamus, who entertains a god named Theuth.[1] Theuth was an inventor of many things including numbers, calculation, astronomy and writing. As Theuth displays his inventions to King Thamus, he describes the invention of writing as something that will "improve both the wisdom and the memory." Thamus replies:

Theuth, my paragon of inventors . . . the discoverer of an art is not the best judge of the good or harm which will accrue to those who practice it. So it is in this case; you, who are the father of writing, have out of fondness for your offspring attributed to it quite the opposite of its real function. Those who acquire it will cease to exercise their memory and become forgetful; they will rely on writing to bring things to their remembrance by external signs instead of on their own internal resources. What you have discovered is a receipt for recollection, not for memory. And as for wisdom, your pupils will have the reputation for it without the reality: they will receive a quantity of information without proper instruction, and in consequence be thought very knowledgeable when they are for the most part quite ignorant. And because they are filled with the conceit of wisdom instead of real wisdom they will be a burden to society.[2]

---

[1]Neil Postman, *Technopoly: The Surrender of Culture to Technology* (New York: Vintage Books, 1993), p. 3.
[2]Plato, *Phaedrus and the Seventh and Eighth Letters*, trans. Walter Hamilton (New York:

Postman goes on to discuss the notion that every new technology has good and bad effects. He warns against the "one-eyed prophets" who only see the benefits of new technologies without imagining their downsides.[3]

Christians with gifts to understand computer technology have a responsibility to discern new developments and engage them. We need to be on guard against being "one-eyed prophets" and to discern the good as well as the potentially harmful aspects embedded in new technologies. The rapid growth and adoption of technology has had a tremendous impact on our culture, and Christians are not immune to its influence. The harmful side effects of some new technologies are not always obvious, even to Christians who work with technology. In fact, there is truth in Thamus's word to Theuth that "the discoverer of an art is not the best judge of the good or harm which will accrue to those who practice it." Generally, whenever we stray from God's intents, sooner or later there are consequences, and this is also the case with technology. For this reason, we need to remain vigilant to the reality of creational norms and steer a correcting course whenever necessary. We need to cultivate a sense of humility as we often muddle through the unexpected issues that arise with new digital technologies. In particular, people who are fluent with computer technology need to be on guard so that we are not "filled with the conceit of wisdom instead of real wisdom."

THE FALL

As discussed in the previous chapter, God made all things good. Unfortunately, near the beginning, the disobedience of humankind led to the fall, which has affected all of creation (see Genesis 3:16-19). The job of humankind was to "till the soil" and look after the earth, but as a result of the fall, the ground became "cursed." Consequently, the ground produced "thorns and thistles," making work a drudgery and much more difficult. These "thorns and thistles" are not limited to agricultural work; the curse extends to all types of work and all creation, including computing and technology. In Romans 8 we read that the "creation was subjected to frustration" and "the whole creation has been groaning as in the pains of childbirth right up to the present time" (Romans 8:20, 22). Exactly how the physical

---

Penguin, 1973), pp. 96-97.
[3]Postman, *Technopoly*, p. 3.

aspects of computation have been "cursed" and "subjected to frustration" is difficult to know. Egbert Schuurman writes, "It is especially difficult—and perhaps impossible—to discern to what extent the pristine creation is present in nature as it now exists."[4] In a perfect world, would computer components be free of decay and network cables free of noise? Questions of this sort are necessarily a matter of speculation.[5]

But computer technology is not just about physical devices; it is a human cultural activity (as we defined it earlier). Hence, sinful human beings misdirect technology in many different ways. Areas in which we exercise freedom and responsibility in shaping technology are tainted by human sin. We go against God's commands and disregard creational norms, leading to distortions and misdirections of technology. Some examples include computer fraud, disregard for privacy, malicious software (such as viruses and worms), cyberbullying and pornographic websites. Computers can also contribute to environmental degradation with improper handling of electronic waste. Evil can be defined as the "spoiling of shalom" and "any deviation from the way God wants things to be."[6]

Nevertheless, it is important to distinguish sin from the good creation and the notion of structure and direction. Albert Wolters refers to *structure* as "the order of creation, to the constant creational constitution of any thing." In contrast, Wolters describes *direction* as "the distortion or perversion of creation through the fall."[7] Sin attaches itself to creation like a parasite.[8] The good possibilities that computer technology brings to creation are intertwined with the effects of sin. The web is useful for communicating and disseminating truthful information; on the other hand, online gambling and pornographic websites are destructive. Email and social networking can shrink the distances between people; but compulsive computer use leads to loss of authentic human contact. Useful software can free us

---

[4]Egbert Schuurman, *Technology and the Future: A Philosophical Challenge* (Toronto: Wedge Publishing, 1980), p. 375.
[5]Joel C. Adams, "Computing Technology: Created, Fallen, In Need of Redemption?" paper presented at the Conference on Christian Scholarship, For What? (Grand Rapids: Calvin College, 2001), p. 2. Available at http://cs.calvin.edu/p/christian_scholarship.
[6]Cornelius Plantinga, *Engaging God's World: A Christian Vision of Faith, Learning, and Living* (Grand Rapids: Eerdmans, 2002), p. 51.
[7]Albert M. Wolters, *Creation Regained* (Grand Rapids: Eerdmans, 1985), p. 49.
[8]Ibid., p. 48.

from countless mundane tasks; but malicious software can cause harm. Nevertheless, the structure of things in creation continues despite their misdirection; bad software is still software. But misdirected technology reminds us of the reality of the fall and how things can be distorted or perverted. Like other things in creation, technology can be directed in obedience or disobedience to God's law.[9] Clearly, technology not only amplifies the potential for greater good but also for greater harm.

Although evil contaminates everything, the image in Romans 8 is one of the "pains of childbirth," which implies that something new is coming! There is a hope of restoration that will finally be realized in the new heaven and earth. In the meantime, we already see indications of God's grace as he restrains the devastating effects of sin in the world. God limits the spread of corruption and preserves the possibility for order and civilization. We can still enjoy the goodness in creation, including the goodness in computer technology.

The preservation of goodness amid a fallen world is not just reserved for Christians. God pours out gifts on believers and unbelievers alike, a concept sometimes referred to as *common grace*. The rain is sent to both the righteous and the unrighteous, and so are the benefits of modern technology (see Matthew 5:45). Common grace can be described as "the goodness of God shown to all, regardless of faith, consisting in natural blessings, restraint of corruption . . . and a host of civilizing and humanizing impulses, patterns, and traditions."[10] In fact, we see evidence of this in useful software and helpful technical contributions that originate from believers and unbelievers alike. The early church father Augustine talked about the treasures of the Egyptians that the people of Israel took with them, and likened them to the good things in culture developed by unbelievers. Augustine suggests that unbelievers can uncover "gold and silver" dug up from "certain mines of divine Providence" that Christians can also take and use.[11] Even though people do not acknowledge the Creator, they still benefit from the goodness of God shown to all.

The first people mentioned in Genesis who were on the forefront of technological developments were all in the line of Cain.[12] It was Cain, cursed by

---

[9]Ibid., p. 49.

[10]Plantinga, *Engaging God's World*, p. 59.

[11]Augustine, *On Christian Doctrine*, trans. D. W. Robertson Jr. (New York: Liberal Arts Press, 1958), p. 75.

[12]For an interesting discussion of this, refer to Jacques Ellul, *The Meaning of the City* (Grand

God for murdering his brother, who became the builder of the first city mentioned in the Bible, naming it after his son Enoch. From the line of Cain came Jubal, the father of all who play the harp and flute. Another son in the line of Cain was Tubal-Cain, who was the first to work with tools of bronze and iron (see Genesis 4:21-22). From the line of Ham, the son of Noah who was also cursed, came Nimrod. Nimrod was a mighty warrior who built many cities, including Babylon and Nineveh, whose reputations are illustrated later in Scripture.

As time went on, people like the Egyptians, Babylonians and Romans were better known for their technological advancements than the people of Israel were. For instance, we read that in the time of Saul, there were no blacksmiths in the land of Israel and they had to go to the Philistines to have their plowshares, mattocks, axes and sickles sharpened (see 1 Samuel 13:19-20). Even Solomon, who built the temple, had to hire the services of Huram from Tyre, who was "a skilled craftsman in bronze" (1 Kings 7:14). Today, many accomplished engineers and computer scientists are people who do not hold to the Christian faith (although there are exceptions). Dutch theologian Abraham Kuyper suggested that God may have imparted some gifts, such as art, to the descendants of Cain so that they may have some "testimony of the divine bounty."[13] Regardless of the faith commitments of those who discover new technologies, we are all working with the bounty of God's creation.

THE TOWER OF BABEL

Genesis also includes the familiar story of the tower of Babel. This story offers an interesting insight into the use of technology. After the flood, God blessed Noah and his sons by saying, "Be fruitful and increase in number and fill the earth" (Genesis 9:1). This underscores that the cultural mandate remained unchanged and that God's intention for humankind and his world remain the same. Genesis 10 continues by recounting the descendants of Noah and how they began to fill the earth a second time. As people multiplied and spread, this was accompanied by cultural and technological

---

Rapids: Eerdmans, 1973).

[13]Abraham Kuyper, *Christianity as a Life-System* (Lexington, KY: Christian Studies Center, 1980), pp. 41-42.

progress. People learned to use "brick instead of stone, and tar for mortar" (Genesis 11:3). This new technology enabled people to begin building a city, which is a part of cultural development.

It did not take long for the possibilities enabled by technology to lead humans to trust in themselves and ignore God. In Genesis we read, "Then they said, 'Come, let us build ourselves a city, with a tower that reaches to the heavens, so that we may make a name for ourselves; otherwise we will be scattered over the face of the whole earth'" (Genesis 11:4). The motivation to build the tower was to make a name for themselves: to glorify humans and their technological prowess. They also built the tower so they would not be scattered. This was in direct defiance of the cultural mandate to "fill the earth," a directive that was repeated in the covenant with Noah (Genesis 1:28; 9:1).

The tower was likely a *ziggurat*: a large structure that served as a staircase by which the gods could descend from heaven to bless a city.[14] Ironically, while humans were busy building a tower "that reaches to the heavens," God still had to come down to see it (Genesis 11:5). God says, "If as one people speaking the same language they have begun to do this, then nothing they plan to do will be impossible for them" (Genesis 11:6). At first glance, these words appear to cast doubt on the notion of creational limits. However, this verse has also been translated as "lest nothing they plot to do be beyond them."[15] In other words, their arrogant plans will observe no limits. Job 42:2 states that no plan of God's can be thwarted, underscoring that human beings are not like God despite what they may plot. In essence, this passage from Genesis depicts how human pride ignores limits and how the people defied God with their plans.

God condemns the arrogance of humans and their disregard for limits and decides to put a stop to their proud ambitions. God confuses their language, which leads to humankind scattering over the face of the earth. This was, in fact, God's original plan: for humans to fill the earth. The word for *Babel* sounds like the Hebrew word for *confused*. Thus, rather than being a

---

[14]Craig G. Bartholomew and Michael W. Goheen, *The Drama of Scripture* (Grand Rapids: Baker Academic, 2004), p. 52.
[15]Gordon J. Wenham, *Genesis 1–15*, Word Biblical Commentary 1 (Waco, TX: Word Books, 1987), p. 233.

monument to human accomplishment, the tower of Babel symbolizes how God confused the sinful plans of human pride.

Other references in Scripture describe people who place their trust in human ingenuity. In the second book of Chronicles, we read of King Asa who, in his sickness, did not seek help from the Lord but only from physicians (see 2 Chronicles 16:12). The warning against trusting in human power is also found in the Psalms. Psalm 20:7 describes those who place their trust in chariots and horses instead of the Lord. Psalm 33:16-19 states that no king is saved by the size of his army, and no warrior by his strength, but rather that it is the Lord who delivers those who trust in him. Psalm 127:1 tells us that unless the Lord builds a house, its workers labor in vain.

The tendency to place our trust in human ingenuity is also apparent in our modern times. Sinful pride and trust in technology lead some to disregard limits. Speculation about emulating brain functions and uploading minds into computers leads some to believe that computer technology will eventually free humans from the limits of their mortality.[16] Virtual reality can promote a sense of power as users exercise control over simulated worlds. The growth of large databases and advanced search engines can foster a sense of having all knowledge at our fingertips. When such activities lead to pride, forgetting God and ignoring limits, they become like building a modern-day tower of Babel.

Accordingly, modern Christians working with computer technology should proceed in humility, even as they uncover the powerful possibilities in creation. In Psalm 8, David considers the heavens and the works of God's hands, and wonders, "What is mankind that you are mindful of them?" (Psalm 8:4). We should have a similar posture. As we work with computer technology, we need to cultivate epistemological humility, realizing that our scientific and technical knowledge is limited. Charles Adams describes epistemological humility as "a posture of appropriate servanthood and creatureliness with respect to our relationship with God and the non-human creation."[17]

---

[16]Ray Kurzweil, *The Singularity Is Near: When Humans Transcend Biology* (New York: Penguin, 2005), p. 325.

[17]Charles Adams, "Galileo, Biotechnology, and Epistemological Humility: Moving Stewardship Beyond the Development-Conservation Debate," *Pro Rege* 35, no. 3 (March 2007): 12.

TECHNICISM AND IDOLATRY

The Enlightenment was accompanied by a growing trust in the power of science and technology. The philosopher Hans Jonas has described this attitude as follows: "To become ever more masters of the world, to advance from power to power, even if only collectively and perhaps no longer by choice, can now be seen to be the chief vocation of man."[18] Secular Western thought replaced trust in God with a trust in human reason. Francis Bacon coined the phrase "knowledge is power," which was taken by many to mean that the value of knowledge lies in its ability to alter one's circumstances to help one attain power. Christian philosopher Nicholas Wolterstorff has called this motivation the "Baconian justification" for the pursuit of knowledge, one that seeks power and autonomy.[19] This widespread belief sees technology as rescuer of the human condition and imagines that technology will eventually solve all our problems. This faith and trust in the power of technology is called *technicism*. Egbert Schuurman describes technicism as "the pretension of humans, as self-declared lords and masters using the scientific-technical method of control, to bend all of reality to their will in order to solve all problems, old and new, and to guarantee increasing material prosperity and progress."[20]

In other words, like the tower of Babel, technicism arises when a culture replaces God with a sense of autonomy and a reliance on technology. This is nothing short of a religion, albeit a false one. Technicism is marked by three key beliefs. First is the belief that the development of increasingly complex objects is inevitable; progress cannot be stopped. Second is the belief that all technological progress will improve the conditions of humankind. Third is the belief that even if technical change brings problems, there will be technical solutions to solve those issues.[21] Christians need to recognize that these beliefs amount to a form of ultimate concern and ultimate trust—in other words, a religion. This trust is a modern form of

---

[18]Hans Jonas, "Toward a Philosophy of Technology," *The Hastings Center Report* 9, no. 1 (February 1979): 38.

[19]Nicholas Wolterstorff, *Reason Within the Bounds of Religion* (Grand Rapids: Eerdmans, 1999), p. 124.

[20]Egbert Schuurman, *Faith and Hope in Technology*, trans. John Vriend (Toronto: Clements Publishing, 2003), p. 69.

[21]Stephen V. Monsma, ed., *Responsible Technology* (Grand Rapids: Eerdmans, 1986), p. 50.

idolatry, something that occurs whenever people substitute their trust in the Creator with a trust in created things (see Romans 1:25).

Although we no longer fashion idols out of wood and stone, as John Calvin noted, "man's nature, so to speak, is a perpetual factory of idols."[22] For some, technology is seen as the route to a better life and greater prosperity, and it gradually takes on a life of its own. There is a role reversal that occurs with idolatry, when the maker of an idol becomes dependent on the idol.[23] The psalmist also warns us regarding idols that "those who make them will be like them, and so will all who trust in them" (Psalm 115:8). An undiscerning trust in digital technology will gradually mold us into patterns of thought that mirror that of a computer. Marshall McLuhan writes, "We become what we behold."[24] Once again, we may shape our machines, but they will also shape us.

Closely related to technicism is *informationism*, which is a "faith in the collection and dissemination of information as a route to social progress and personal happiness."[25] Its proponents suggest that universal access to information and the Internet will bring enlightenment and make the world a better place. Nicholas Negroponte, the founder of the Massachusetts Institute of Technology's Media Lab, wrote a book shortly after the web was invented called *Being Digital*. In it he remarks, "A new generation is emerging from the digital landscape free of many of the old prejudices. These kids are released from the limitation of geographic proximity as the sole basis of friendship, collaboration, play, and neighborhood. Digital technology can be a natural force drawing people into greater world harmony."[26]

This view is not new. In the mid-nineteenth century, the advent of the transatlantic telegraph brought many optimistic predictions, including some suggestions that it could lead to world peace. Negroponte's optimism about digital technology echoes things that were written about the telegraph. Already in 1858, authors Briggs and Maverick wrote in *The Story of the Telegraph*, "How potent a power, then, is the telegraphic destined to become in

---

[22]John Calvin, *Institutes of the Christian Religion*, vol. 1, ed. John T. McNeill, trans. Ford Lewis Battles (Philadelphia: Westminster Press, 1960) (Bk. 1, 11.8), p. 108.

[23]Bob Goudzwaard, *Idols of Our Time* (Downers Grove, IL: InterVarsity Press, 1984), p. 22.

[24]Marshall McLuhan, *Understanding Media: The Extensions of Man* (New York: McGraw Hill, 1964), p. 19.

[25]Quentin J. Schultze, *Habits of the High-Tech Heart: Living Virtuously in the Information Age* (Grand Rapids: Baker Academic, 2002), p. 21.

[26]Nicholas Negroponte, *Being Digital* (New York: Knopf, 1995), p. 230.

the civilization of the world! This binds together by a vital cord all the nations of the earth. It is impossible that old prejudices and hostilities should longer exist, while such an instrument has been created for an exchange of thought between all the nations of the earth."[27] A popular slogan suggested that the global telegraph would "make muskets into candlesticks."[28] The mistaken notion that peace and prosperity will come with communications technology illustrates a kind of informationism. It is misguided to think that worldwide access to communication and information technology will ever solve the problems of the human condition.

There are other beliefs related to technicism. *Scientism* claims that human reason can provide complete understanding of humanity and nature. C. Stephen Evans describes scientism as "the belief that all truth is scientific truth and that the sciences give us our best shot at knowing 'how things really are.'"[29] This view does not recognize that our problems are due to sin, not just a lack of knowledge. Technicism is closely related to scientism because science provides the knowledge that can be used to control nature and achieve power.[30]

Scientism drives technicism, which, in turn, feeds *consumerism*. The premise of consumerism is a belief that people can find happiness through purchasing and consuming material goods. Technology has played a significant role in the spread of consumerism. Computer technology and automation have enabled products to be produced in large quantities and at low cost. Raw materials can be harvested, processed, shipped and packaged at an astonishing rate. Short technology life cycles, driven by Moore's Law, and the rapid pace of change mean that many of these products are soon discarded in exchange for even newer fads and features. Technology has itself become a product, used to entice one to continually "upgrade," "update" and "modernize." The prices of electronic gadgets often do not reflect their true cost, including the human and environmental costs.[31] Fur-

---

[27]Charles F. Briggs and Augustus Maverick, *The Story of the Telegraph* (New York: Rudd & Carleton, 1858), p. 22.
[28]Tom Standage, *The Victorian Internet* (New York: Walker & Company, 2007), p. 83.
[29]C. Stephen Evans, *Preserving the Person: A Look at the Human Sciences* (Vancouver, BC: Regent College Publishing, 2002), p. 18.
[30]Brian J. Walsh and J. Richard Middleton, *The Transforming Vision: Shaping a Christian World View* (Downers Grove, IL: InterVarsity Press, 1984), p. 133.
[31]The true environmental cost of electronic devices can be better appreciated using approaches

thermore, technology also feeds consumerism by providing a variety of different electronic media so that a continuous stream of advertising can lure us into purchasing more. More recently, web technology has enabled advertisers to track, profile and target people more effectively than ever before.[32] We are surrounded with advertisements (both online and in the "real world"), and we are often unaware of the extent to which they steadily make us covet more. At its core, consumerism is yet another form of idolatry, with its hollow promise to deliver satisfaction apart from God.

To elevate one aspect of creation gives shape to a worldview through which everything else is understood. Jacques Ellul identifies technique as the view that all the world's problems can be addressed by rational and efficient methods. In this case, the elevation of efficiency and rational methods give shape to a particular worldview. Karl Marx explained all of society in terms of the history of class struggle. Others have elevated nationhood, material prosperity and security into ideologies that take on ultimate concern, resulting in idolatry.[33]

If technology has the tendency to become an idol in our age, should Christians avoid technology altogether? In 1 Corinthians 8, Paul addresses a similar issue. Should leftover meat be rejected if it had been sacrificed to idols?[34] His answer is that leftover meat need not be rejected but could be served to Christians or perhaps even sold in the public meat market. Paul argues that "'an idol is nothing at all in the world' and that 'There is no God but one'" (1 Corinthians 8:4). Later, he states, "Eat anything sold in the meat market without raising questions of conscience, for 'The earth is the Lord's, and everything in it'" (1 Corinthians 10:25-26). Clearly, Christians are free to enjoy God's creation, even if others worship it. The only limitation Paul gives is to show concern for those who are not mature in the faith. We need not avoid computer technology; instead, we need to demonstrate its rightful place and its normative use.

---

such as Life-Cycle Analysis (LCA), which assesses the environmental impact of a product from cradle to grave, including raw material extraction, manufacture, distribution, use, maintenance and disposal or recycling.

[32]Stephen Baker, *The Numerati* (New York: Houghton Mifflin Harcourt, 2008), pp. 41-66.

[33]Bob Goudzwaard's *Idols of Our Time* does a good job of exploring these ideologies.

[34]Richard J. Mouw, *When the Kings Come Marching In* (Grand Rapids: Eerdmans, 2002), p. 33.

TECHNOLOGY AS A RESULT OF THE FALL?

The connection between technology and the fall prompts a puzzling question: is technology itself a result of the fall? Some might suggest that a perfect world would have no need for technology. Jacques Ellul describes the ideal state of creation as follows: "No cultivation was necessary, no care to add, no grafting, no labor, no anxiety. Creation spontaneously gave what man needed."[35] In a "world where there was no necessity," Ellul questions what possible purpose there would be for technique. Ellul suggests, "Thus, no matter what attitude one takes toward technique, it can only be perceived as a phenomenon of the fall; it has nothing to do with the order of creation; it by no means results from the vocation of Adam desired by God. It is necessarily of the situation of the fallen Adam."[36]

There are certainly distortions in how technology and efficiency have shaped human activities, but is technology a result of the fall? It is helpful once again to distinguish between creational structure and its direction.[37] God continues to uphold the structures of creation, but sin has corrupted the direction of technology. Technology and rational methods are part of the structure of creation; however, they can be absolutized or misdirected. In the previous chapter we asserted that technology is part of the latent potential in creation. Technology is not a result of the fall; rather, it is a human cultural activity that is part of the possibilities in creation. In fact, the cultural mandate was given by God to humankind before the fall even occurred. Therefore, Christians have a responsibility to discern the technological possibilities in creation and apply them in ways that honor God. Our definition of computer technology underscores this by highlighting that it is a cultural activity in which humans exercise freedom and responsibility in response to God.

ANTINORMATIVE TECHNOLOGY

As stated earlier, things in creation have both a structure and a direction. This is true for institutions like schools, for cultural things like art and also

---

[35]Jacques Ellul, "Technique and the Opening Chapters of Genesis," in *Theology and Technology: Essays in Christian Analysis and Exegesis*, ed. Carl Mitcham and Jim Grote (Lanham, MD: University Press of America, 1984), p. 126.
[36]Ibid., p. 135.
[37]Wolters, *Creation Regained*, p. 49.

for technology.[38] Computer technology is not neutral; it can either be directed in ways that comport well with God's intentions for his world or in rebellious ways. As humans, we have freedom and responsibility in response to God as to how we direct our technical activities.

At this point it is helpful to reach into our toolbox and refer to the modal aspects introduced in the previous chapter. Recall that the earlier modal aspects, such as the numeric and kinetic modalities, are described by various laws, whereas the later modalities have norms associated with them. Although creational laws are fixed, humans have considerable freedom (and hence responsibility) when it comes to the normative aspects. We may choose to ignore creational norms; however, we do so at our peril. Ignoring normative principles goes against the fabric of creation and entails negative consequences. Christian economist Bob Goudzwaard describes this as follows: "If man and society ignore genuine norms . . . they are bound to experience the destructive effects of such neglect. This is not, therefore, a mysterious *fate* which strikes us; rather, it is a *judgment* which men and society bring upon themselves. . . . To ignore given norms out of an *a priori* illusion of autonomy only seems to afford freedom, but in the long run it removes genuine freedom. . . . Genuine laws or norms are pointers that guide us along safe and passable roads. Apart from norms our paths run amok."[39]

The consequences of ignoring norms act as warnings that things are not the way they are supposed to be and that we ought to change our behavior. Another way of stating this is that creational norms act like elastic bands that we can push against but that sooner or later will have a way of pushing back. These norms are also active when we work with computer technology. To avoid running amok, the development and use of computer technology should be guided from the start with sensitivity toward creational norms. It is because of sin that computer technology is often used and developed in antinormative ways.

The effects of ignoring certain norms can be observed in the case of compulsive computer use. We were created as physical beings meant to live in

---

[38]Ibid.
[39]Bob Goudzwaard, *Capitalism and Progress: A Diagnosis of Western Society* (Grand Rapids: Eerdmans, 1979), p. 243.

community; when we spend excessive amounts time gaming or surfing, there are consequences. Constant updates, messages and notifications lead to a state of continuous partial attention. Unrelenting time spent online can leave people fatigued and in a "digital fog."[40] Norms also apply to human communications. Whereas email and social networking can shrink distances, they are no substitute for face-to-face communications. The nuances of body language, facial expressions and intonation cannot be fully transmitted via email—not even with the clever use of emoticons.[41] Many important aspects of communication are lost when transmitted solely via electronic media. Occasionally this can lead to misinterpretations or misunderstandings. Moral norms are ignored in the case of online pornography and gambling. Norms also extend to areas of ethics and law, such as the need to respect intellectual property and copyrights. There are also aesthetic norms that apply in computer technology, as is the case in the design of graphical user interfaces, websites and ergonomics.

Furthermore, there are norms that apply to the economic and project management aspects of computer technology. A type of tunnel vision occurs when technology is driven narrowly by monetary or economic considerations. A technical worldview directs things toward efficiency at the expense of many other considerations. Egbert Schuurman writes, "The norms that follow from the values of the technological world picture are effectiveness, standardization, efficiency, success, reliability, and maximum profit, with little or no attention given to the cost to humanity, society, the environment, and nature."[42] The Dilbert comic strip illustrates this in an amusing way with a hapless engineer, Dilbert, who works in a corporation driven strictly by marketing and monetary considerations and the absurd situations, doomed projects, frustration and defective products that ensue.

Furthermore, reducing people to their economic functioning fails to recognize their value as human beings. This attitude is evident when electronics and computers are manufactured in developing countries without

---

[40]Gary Small, *iBrain: Surviving the Technological Alteration of the Modern Mind* (New York: William Morrow, 2008), p. 19.

[41]An emoticon is a text representation of the writer's face or mood using combinations of keyboard characters.

[42]Egbert Schuurman, *The Technological World Picture and an Ethics of Responsibility* (Sioux Center, IA: Dordt College Press, 2005), p. 22.

due care for workers and working conditions. Focusing on profits and ignoring concerns like ergonomics can lead to harm for the end users. Software projects that ignore sound management practices, that require grueling hours from programmers and that are not planned properly often fail. Projects like these often lead to personal stress for programmers and are sometimes referred to as a "death march."[43] In short, ignoring norms will sooner or later lead to consequences.

Over time, people have identified patterns of poor practices called *antipatterns*.[44] In a book titled *AntiPatterns*, the authors identify some of the root causes of antipatterns including apathy, avarice, sloth, ignorance and pride.[45] Although some antipatterns deal with technical issues, many of them can be attributed to antinormative practices. Whether or not people acknowledge it, they are bumping up against creational norms.

Amid rapid technological advancements, we should maintain a posture of humility, understanding that in our fallen state we see only dimly (see 1 Corinthians 13:12). The job of humbly discerning and wrestling with the normative principles surrounding computer technology is hard work. For this reason, it is prudent to temper the pace at which new technologies are developed to ensure adequate time for reflection on their use and consequences. Even so, there will be missteps that will require course corrections. The job of working this out is not just an individual task but also the job of the larger Christian community.[46] These norms will be explored in further detail in chapter 4.

COMPUTER BUGS

In 1945, one of the first electronic computers at Harvard mysteriously halted. After tracing through the machine, the fault was located: a moth had been crushed between some relay contacts. The computer literally had a

---

[43]Edward Yourdon, *Death March: The Complete Software Developer's Guide to Surviving 'Mission Impossible' Projects* (Upper Saddle River, NJ: Prentice Hall, 1999).

[44]This is in contrast to *design patterns*, which are good solutions to common problems in software design.

[45]William J. Brown et al., *AntiPatterns: Refactoring Software, Architectures, and Projects in Crisis* (San Francisco: Wiley, 1998), pp. 19-24.

[46]Some organizations, such as the Association of Christians in the Mathematical Sciences (ACMS), are examples of communities of Christians who are working to discern a Christian approach to mathematics and computer science. See www.acmsonline.org.

bug. Since then, *computer bug* has become a generic term referring to software errors that continue to plague modern computers.

Computer bugs can be compared to thorns and thistles, which provide an apt metaphor for many of the problems that arise in computer work and research. Weeds inevitably appear in computer work, just as they do in a garden. The necessary process of testing and debugging is nothing less than "weeding" out a computer program, and it is just as time-consuming and arduous as weeding a real garden. Furthermore, weeding a garden is a never-ending task because new weeds can always sprout.

Likewise, in computing, we can never be sure that a computer program is totally bug-free. Besides the software, we cannot be sure the original program specifications are even correct.[47] Edsger Dijkstra observed, "Testing can be used to show the presence of bugs, but never to show their absence."[48] Likewise in a garden, a visual inspection will not uncover the weeds that may be lying just beneath the surface, waiting to sprout. Sometimes removing a bug is like removing a weed; you cannot always be certain that it has been completely removed at the root. Previously "fixed" bugs will sometimes reappear just as a weed can reappear if a portion of the root remains after it is pulled. Debugging requires isolating and fixing the source of the bug to get at the "root" of the problem. In brief, gardening and computing appear to have some striking similarities.

The area of software development is also fraught with difficulties. Frederick Brooks likens large software projects to a tar pit when he says, "No scene from prehistory is quite so vivid as that of the mortal struggles of great beasts in the tar pits. . . . The fiercer the struggle, the more entangling the tar, and no beast is so strong or so skillful but that he ultimately sinks. Large-system programming has over the past decade been such a tar pit. . . . Large and small, massive or wiry, team after team has become entangled in the tar."[49] The tar pit is also an apt analogy to for the nu-

---

[47]An example is the accident of Lufthansa Flight 2904 in 1993, which went off the runway when software failed to activate the thrust reverse system. This was due to a software specification that did not account for unexpected circumstances. See Frederick P. Brooks, *The Design of Design* (Boston: Addison-Wesley, 2010), p. 110, n. 6.

[48]Edsger W. Dijkstra, "Notes on Structured Programming," 2nd ed. (technical report, EWD249, Technical University Eindhoven, Eindhoven, Holland, April 1970), p. 7.

[49]Frederick P. Brooks, *The Mythical Man-Month: Essays in Software Engineering* (San Francisco: Wiley, 1995), p. 4.

merous pitfalls that can plague the software development process.

Separating the creational aspects of computer programming (good structure) from the effects of the fall (bad direction) can be difficult to discern. For instance, the inherent complexity of computer programming is likely a part of creation and not a result of the fall. Likewise, the iterative process in which computer programs are written, tested and updated may just be intrinsic to the activity of computer programming. Is bug-fixing simply part of the activity of writing complex programs, or is it a result of the fall? This is difficult to discern. Nevertheless, it can be said that the drudgery and harmful consequences of bugs and failed computer projects are certainly a result of the fall.

To make software more robust, it is prudent to anticipate user errors that may occur with computer systems. The book *The Design of Everyday Things* includes a section titled "Designing for Error," which includes stories of user errors and some thoughts on minimizing the occurrence and severity of errors. Proper design considerations can help minimize the causes of error, make it possible to reverse some errors and make errors easier to discover.[50] Errors cannot be eliminated entirely, but good design practices can help reduce them.

Despite the thorns and thistles, technology is still part of the original goodness of creation. Even in the biblical account of the fall, we see a glimpse of how technology might be used. Immediately after the fall, Adam and Eve clothed themselves with leaves. After God found them, however, he made them more durable clothes from animal skins (see Genesis 3:21). This demonstrates God's grace and the aid that the earth's resources offer humans in their fallen state. In other words, technology can be used to "push back" some of the effects of the fall. This is particularly evident in fields such as medicine, where technology is used to treat various diseases. Egbert Schuurman suggests that meaningful functions for technology will include "emancipating body and mind from toil and from drudgery, repelling the onslaughts of nature, providing for man's material needs, and conquering diseases."[51]

---

[50]Examples of design that makes it possible to reverse errors are software applications that include an "undo" button. If only such a button were possible in other areas of life! See Donald A. Norman, *The Design of Everyday Things* (New York: Basic Books, 1988), p. 131.

[51]Egbert Schuurman, *Reflections on the Technological Society* (Toronto: Wedge Publishing, 1977), p. 21.

Nevertheless, in the same breath we must acknowledge that because technology is itself affected by the fall, it will never rescue or restore us completely. In the words of Nicholas Wolterstorff, "Technology does make possible advance toward shalom; progress in mastery of the world can bring shalom nearer. But the limits of technology must also be acknowledged: technology is entirely incapable of bringing about shalom between ourselves and God, and it is only scarcely capable of bringing about the love of self and neighbor."[52] Computer technology is like any other aspect of creation: it can be directed for good, but it remains fallen and limited.

CONCLUSION

The disobedience of humankind led to the fall, which has affected all human activities, including computer technology. The tower of Babel tells the story of misplaced trust in technology and the consequences of human pride. The modern equivalent to a tower of Babel mentality is called technicism, and it occurs when people replace trust in God with a reliance on the possibilities of modern technology. The human heart still fashions golden calves, and technology is one of the idols of our times. Sin has resulted in misdirected and antinormative technology, which has led to numerous consequences. But thankfully, this is not the end of the story.

---

[52]Nicholas Wolterstorff, *Until Justice and Peace Embrace* (Grand Rapids: Eerdmans, 1983), p. 71.

4

Redemption and Responsible
Computer Technology

In the end, God will come to fix his world and make it altogether good again.
In between, his children are to go into the world and create
some imperfect models of the good world to come.

LEWIS SMEDES, *MY GOD AND I*

■■■

Despite the extent to which sin has stained our world, there is reason for hope. God has not abandoned his creation to despair, and he does not want us to abandon it either. Although the world is broken by sin, we can have hope that "there is still enough goodness in the world to make it both fixable and worth fixing."[1]

The atoning work of Jesus Christ is central to God's plan to restore his creation. Colossians 1 provides a clear account of the centrality of Jesus Christ and the cross.

> For in him all things were created: things in heaven and on earth, visible and invisible, whether thrones or powers or rulers or authorities; all things have been created through him and for him. He is before all things, and in him all things hold together. And he is the head of the body, the church; he is the beginning and the firstborn from among the dead, so that in everything he might have the supremacy. For God was pleased to have all his fullness dwell in him, and through him to reconcile to himself all things, whether things on earth or things in heaven, by making peace through his blood, shed on the cross. (Colossians 1:16-20)

---

[1]Lewis Smedes, *My God and I* (Grand Rapids: Eerdmans, 2003), p. 59.

We read that Christ is central in creation and that it is "through him" that all things were created. This fact is highlighted at the beginning of the Gospel of John, where we are told that it was through him that all things were made (see John 1:3). Furthermore, it is Christ who sustains all things; it is "in him" that all things hold together. There is also a teleological order or purpose to creation, since all things were created "for him." It is through Christ's death on the cross that God is reconciling all things to himself. It is through Christ that all things are being redeemed.

The repeated words "all things" highlight that salvation is comprehensive in scope; it is about more than personal salvation. Because sin is so comprehensive, redemption is equally cosmic in scope. In the words of the familiar Christmas hymn, "He comes to make his blessings flow, far as the curse is found."[2] Salvation is not only about forgiveness of individual sins, but about restoring all of creation to its intended state. Oliver O'Donovan writes that "the resurrection of Christ directs our attention back to the creation which it vindicates."[3] Colin Gunton underscores the connection of redemption to creation when he writes, "What we call redemption is not a new end, but the achievement of the original purpose of creation."[4] Thus Christ's redemption involves the restoration of all things in creation, and we are called to participate in this process.

The scope of "all things" includes computer technology, but does computer technology need redemption? What does a Christian approach to computer technology look like? Abraham Kuyper wrote about the difference between believers and nonbelievers and how belief transforms society. Kuyper stated that there are essentially two kinds of people: those transformed by God and everyone else. He continues this reasoning as follows: "but one is inwardly different from the other, and consequently feels a different content rising from his consciousness; thus they face the cosmos from different points of view, and are impelled by different impulses. And the fact that there are two kinds of *people* occasions of necessity the fact of two kinds of human *life* and

---

[2]From "Joy to the World" by Isaac Watts.
[3]Oliver O'Donovan, *Resurrection and Moral Order: An Outline for Evangelical Ethics* (Downers Grove, IL: InterVarsity Press, 1986), p. 31.
[4]Colin E. Gunton, *Christ and Creation* (Eugene, OR: Wipf & Stock, 1992), p. 94.

consciousness of life, and of two kinds of *science*."[5]

This idea raises many interesting questions. Does the Christian faith result in a "new kind" of computer technology? Should we build Christian computers or Christian operating systems?[6] Should we be making Christian software, just as some have formed Christian schools and Christian farmers associations? What would a Christian word processor look like, and what about the possibility of a Christian Internet?

Suppose two programmers—one a Christian and one an atheist—set out to write a computer program. Both use the same programming language, the same compiler and the same operating system; both employ the same software engineering techniques. Can the end user discern the religious convictions of the programmer? If not, what difference does faith make to our work in computer science?

Sietze Buning wrote a poem titled "Calvinist Farming," in which he reflected on how some past generations of Calvinist farmers were distinctive in the way they dressed and farmed. They wore "neckties with their bib-overalls . . . a touch of glory with their humility."[7] They planted their corn in straight rows running east-west or north-south in checkerboard patterns, in a predetermined fashion that reflected their views on divine election. No Calvinist followed the land's contours and planted their rows against the slope of the land, like farmers in surrounding counties did. The placement of corn stalks in contour farming was not predetermined, but more like free will. Contour farmers demonstrated a frivolous attitude toward the doctrine of election, as evidenced by the way they planted. The goal of the Calvinist farmers was to farm on biblical principles: decently and in good order. Leaving aside the issue of soil erosion, it was possible to discern their faith commitments by the way they farmed.

Is there a "new kind" of engineering or computer science that is distinctive? Like the farmers in Buning's poem, can we program computers in a way that people can discern our faith commitments? Is there a Christian approach to computer technology? Can one's faith genuinely shape the dis-

---

[5] Abraham Kuyper, *Principles of Sacred Theology* (Grand Rapids: Baker, 1980), p. 154.
[6] In fact, there actually is a project, called Ichthux, aimed at building a version of the Linux operating system for Christian users.
[7] Sietze Buning, "Calvinist Farming," in *Purpaleanie and Other Permutations* (Orange City, IA: Middleburg Press, 1978).

cipline of computing without becoming forced or artificial? If faith does make a difference in this area, where do we start?

An important reminder is that our aim is not to be different for its own sake; rather, any differences that arise in our approach to computer technology should be as a consequence of our beliefs. Nicholas Wolterstorff summarizes this notion as follows: "Faithful scholarship will, as a whole, be *distinctive* scholarship; I have no doubt of that. But difference must be a consequence, not an aim. And if at some point the difference is scarcely large enough to justify calling this segment of scholarship a 'different kind of science'—*Christian* science in contrast with competitors which are *non*-Christian—why should that, as such, bother us? Again, isn't *faithful* scholarship enough? Difference is not a condition of fidelity—though, to say it once more, it will often be a *consequence*."[8]

Historian George Marsden has written on the topic of how faith can have a direct bearing on scholarship in *The Outrageous Idea of Christian Scholarship*. In a chapter titled "What Difference Could It Possibly Make?" Marsden identifies at least four ways that faith could make a difference. First, faith can be a motivation to do one's work well. Second, faith can help determine the applications for one's scholarship. Third, faith can shape the questions one asks about one's field of inquiry. Fourth, faith influences how a scholar sees her overall field, its meaning and how it relates to the big picture.[9]

These points are also relevant to the study and use of computer technology. Our faith can motivate us to do our technical work well as a way to be faithful stewards and to show love for our neighbors. Our faith can also help us seek fruitful areas in which to use computers and avoid areas that would bring harm. Faith can shape the questions we ask about computer technology, such as how well it comports with creational norms. Finally, our faith helps us see computer technology in the context of a biblical worldview.

The Bible gives us the big picture, but how do we apply this to computer technology? There are no references for the word *computer* in a Bible concordance. Attempts to force-fit verses into situations for which they were

---

[8]Nicholas Wolterstorff, "On Christian Learning," in *Stained Glass: Worldviews and Social Science*, ed. Paul A. Marshall, Sander Griffioen and Richard J. Mouw (Lanham, MD: University Press of America, 1989), p. 70.
[9]George Marsden, *The Outrageous Idea of Christian Scholarship* (New York: Oxford University Press, 1998), pp. 63-64.

never intended leads to biblicism. Nevertheless, God's Word can still guide our decisions regarding the use of computer technology. As John Calvin said, the Scriptures are a guide and teacher that, like "spectacles," help us to see more clearly.[10] Another helpful analogy is that of an automobile, in which the Bible is represented by the engine and the area of application as the wheels. The engine is normally connected to the wheels via some sort of transmission, and in Christian scholarship the transmission can be provided by a Christian worldview. A Christian worldview is based on a biblical framework and is guided by biblical themes and biblical norms.[11] It is a Christian worldview that facilitates connecting the light of Scripture to all areas of study where "the rubber hits the road."[12]

A good place to start in developing a Christian worldview is to look at God's general goal for human existence: shalom. Shalom is often translated as "peace," but it is more than that. Shalom means, in the words of Cornelius Plantinga, a "universal flourishing, wholeness, and delight—a rich state of affairs in which natural needs are satisfied and natural gifts fruitfully employed, all under the arch of God's love. Shalom, in other words, is the way things are supposed to be."[13] Christian hope for shalom begins with Christ, who reconciles all things "through his blood, shed on the cross" (Colossians 1:20). As followers of Jesus Christ, we are called to be agents of shalom.

If shalom is the way things are supposed to be, how do we know how things are supposed to be for computer technology? Some issues are addressed directly in Scripture: we should not, for example, steal from or harm our neighbor. But there are numerous other technology-related issues that the Bible does not specifically address. What does the Bible have to say, for example, about issues such as personal information, privacy, intellectual property and artificial intelligence?

Unfortunately, this concern may lead some to conclude that the Bible cannot guide us in matters relating to computers. This results in a dualistic

---

[10]John Calvin, *Commentaries on the First Book of Moses, called Genesis,* vol. 1, trans. John King (Grand Rapids: Eerdmans, 1948), p. 62.

[11]Sidney Greidanus, "The Use of the Bible in Christian Scholarship," *Christian Scholar's Review* 11, no. 2 (1982): 145.

[12]I am grateful to Al Wolters for introducing me to this analogy in discussions about Christian scholarship.

[13]Cornelius Plantinga, *Engaging God's World: A Christian Vision of Faith, Learning, and Living* (Grand Rapids: Eerdmans, 2002), p. 15.

view of life—believing that some parts of life are "sacred" and others are not. It also perpetuates the notion that some careers are more "spiritual" than others. Some Christians may shun a technical career in favor of a more "religious" career, such as missionary or minister. From this perspective, technicians and engineers are like the modern-day "Marthas" of our world: always busy in the kitchen, looking after the technical details, while others focus on sitting at the feet of Jesus.

But the gospel is not only confined to areas such as church and personal piety. Jesus is Lord of every square inch, and each of us in our small corner is called to work toward shalom. This notion was captured well in the familiar quote from Abraham Kuyper, who declared, "There is not a square inch in the whole domain of our human existence over which Christ, who is sovereign over all, does not cry: 'Mine!'"[14] The notion of the "priesthood of all believers" implies that all of us, whether pastors or programmers, are to work in humble service for our Lord. God's people are called to all sorts of vocations, including the area of computer technology. This message comes across quite clearly in the Bible. In 1 Peter 4:10 we read, "Each of you should use whatever gift you have received to serve others, as faithful stewards of God's grace in its various forms." Paul writes, "So whether you eat or drink or whatever you do, do it all for the glory of God" (1 Corinthians 10:31). Elsewhere we read, "And whatever you do, whether in word or deed, do it all in the name of the Lord Jesus, giving thanks to God the Father through him" (Colossians 3:17). The Christian faith is comprehensive in scope, and it has implications for all areas of life. In the words of theologian Gordon Spykman, "Nothing matters but the kingdom, but because of the kingdom everything matters."[15] A frequently cited quotation often attributed to Karl Barth summarizes it this way: "A Christian is a thinking person who holds the Bible in one hand and the newspaper in the other."[16] We may expand on this thought by stating that a contemporary version of that newspaper could just as well be a tablet computer or smartphone. Each new generation is called to apply

---

[14]Richard J. Mouw, *Abraham Kuyper: A Short and Personal Introduction* (Grand Rapids: Eerdmans, 2011), p. 4.

[15]Gordon J. Spykman, *Reformational Theology: A New Paradigm for Doing Dogmatics* (Grand Rapids: Eerdmans, 1992), p. 266.

[16]This quotation is often attributed to Karl Barth, but he may not have said it. I was not able to find an authoritative source for this quotation.

a biblical worldview to the contemporary issues of their day.

In Proverbs, we are advised, "Trust in the Lord with all your heart and lean not on your own understanding; in all your ways submit to him, and he will make your paths straight" (Proverbs 3:5-6). The metaphor of a path is used, and if we follow God, he will make our paths straight. This is similar to the promise in Psalm 23, where the Lord is likened to a shepherd guiding us along the right paths. Technicism is an example of leaning on our own technical understanding, which leads along the wrong path and to the wrong destination. What is a straight path? It's not the mathematical definition of the shortest path between two points; rather, it is setting our eyes on Jesus and running the race. The path is the line between us and Jesus.

God's Word sets out general guidelines that serve as guardrails to help us safely follow a straight path. There are biblical norms as well as norms in the creation that we need to discern. This is accomplished through prayer, the study of God's Word and the wise counsel of fellow believers. Furthermore, God grants us discernment through the inner working of the Holy Spirit. In Romans we read, "Do not conform to the pattern of this world, but be transformed by the renewing of your mind. Then you will be able to test and approve what God's will is—his good, pleasing and perfect will" (Romans 12:2). We are not alone; we have God's Word and his guidance as we navigate these new paths.

Normative Principles for Technology

In the first chapter we asserted that technology is never neutral; it is value-laden. Technology not only has a structure; it also has a direction. Just because something *can* be done does not imply that it *ought* to be done. We need to discern normative principles that set the values and direction of computer technology. Furthermore, these principles must comport well with God's Word and with the fabric of creation.[17]

Since computing technology is applied in many different areas, it is frequently a multidisciplinary activity. Although computing objects have the technological modality as a foundation, they often have another modality as a destination. For instance, computers used in modern traffic control

---

[17]Stephen V. Monsma, ed., *Responsible Technology* (Grand Rapids: Eerdmans, 1986), p. 61.

involve technical details but have a social purpose. Accordingly, the primary responsibility for such a system should include a traffic expert in cooperation with computer scientists and engineers. Development of software to assist with medical care should be guided by physicians and healthcare experts, in collaboration with computer programmers. Computer scientists and engineers should seek guidance from domain experts during the design process. These experts help ensure that design decisions are not driven primarily by technical considerations but rather with the primary purpose and end user in mind. Many recent software engineering strategies recognize this by specifying continuous customer and stakeholder input and collaboration throughout the development process.[18] In such cases, development of technology becomes an interdisciplinary and communal responsibility.

Domain experts, computer developers and end users need to be guided by norms. These norms begin with God's law, which was summarized by Jesus in this way: to love God above all and to love one's neighbor as oneself (see Matthew 22:37-40). Hence, the overarching normative principle is one of love. Part of how we love God is by faithfully carrying out the cultural mandate and showing respect for his creation. One way to love our neighbor is by opening up, not constricting, the opportunities for them to be the people God wants them to be.[19] Good technology is consistent with a respect for people and in the service of all kinds of life. To follow God's norms is to seek the way God intended things to be and to seek shalom. Goudzwaard describes the purpose of norms as follows: "The purpose of norms is to bring us to life in its fullness by pointing us to paths which safely lead us there. Norms are not straitjackets which squeeze the life out of us . . . the created world is attuned to those norms; it is designed for our willingness to respond to God and each other."[20] And so it is with computer technology; we must discern how to go with the grain of creation rather than against it.

Donald Norman, author of *The Design of Everyday Things*, makes this observation: "If everyday design were ruled by aesthetics, life might be more pleasing to the eye but less comfortable; if ruled by usability, it might be more

---

[18]An example is the Agile Software Development method, which specifies continuous customer involvement.

[19]Monsma, *Responsible Technology*, p. 69.

[20]Bob Goudzwaard, *Capitalism and Progress: A Diagnosis of Western Society* (Grand Rapids: Eerdmans, 1979), pp. 242-43.

comfortable but uglier. If cost or ease of manufacture dominated, products might not be attractive, functional, or durable. Clearly, each consideration has its place. Trouble occurs when one dominates all the others."[21] Considerations such as aesthetics, usability and cost are aspects with which we exercise freedom and responsibility. There are many other aspects to consider as well. For this reason we will once again reach into our toolbox to use the modal aspects introduced in the second chapter. Recall that all the modal aspects come into play when using or designing computer technology. The earlier modal aspects, such as the arithmetic and the physical, involve creational laws that we have no choice but to obey. The later aspects are associated with norms and values, for which humans have a degree of choice and freedom.

Each of the normative aspects ought to be informed by biblical norms. For instance, the juridical aspect is informed by the biblical norm of justice, and the economic aspect is informed by the biblical norm of stewardship. Since the fabric of creation is multifaceted, these norms work together and are not mutually exclusive. In the words of Goudzwaard, "The norms of economic development and those of ethics, the norms of justice and the unfolding of technique, ought never to be played off against each other. Because God's command is undivided, the norms set by him must be seen and observed in their mutual coherence."[22] All of the normative aspects ought to be shaped by the biblical norms of love and care and should contribute to shalom.

Applying the modal aspects to areas related to computing has been explored by author Andrew Basden, who illustrates an "aspectual analysis" of user interfaces.[23] He also provides some examples of visualizing "aspectual normativity" by using bar charts. Basden refers to a concept called the "shalom principle," which states that "if we function well in every aspect then things will go well, but if we function poorly in any aspect, then our success will be jeopardized."[24] In short, it is prudent to keep an eye on all the normative aspects because they are all important.

In the following sections, we will cover each of the last eight normative

---

[21]Donald A. Norman, *The Design of Everyday Things* (New York: Basic Books, 1988), p. 151.
[22]Goudzwaard, *Capitalism and Progress*, p. 65.
[23]Andrew Basden, *Philosophical Frameworks for Understanding Information Systems* (Hershey, PA: IGI Global, 2007), pp. 153-55.
[24]Ibid., pp. 156-58; p. 105.

aspects as a framework to survey contemporary issues in computer technology. These aspects include the historical (cultural), lingual, social, economic, aesthetic, juridical, ethical and faith aspects. Many of the issues that will be touched upon are complex and demand a much more thorough treatment than can be provided here. These norms do not provide pat answers for difficult issues, but rather point a way forward. Responsible computer technology begins with understanding norms and how technology is value-laden. These values will become more apparent as we explore various normative aspects as they relate to computer technology.

HISTORICAL AND CULTURAL NORMS

The historical modality relates to the forming and developing of human culture, and so the associated norm is one of cultural appropriateness.[25] This involves making decisions that take into account the context, users and setting in which computer technology will be used. This norm should be shaped by the biblical norms of love and care. For example, software systems that are introduced into a company or organization should not force people to adapt to the software; rather, the software should accommodate the needs of those who will use it. Other examples arise when particular computer technologies are exported to different countries or cultures where their use may disrupt existing practices or social structures. Cultural appropriateness pertains not only to foreign cultures but should also be considered when introducing computer technology into distinct spheres, such as the school and the church. A technology that adapts to its culture can actually enrich it.[26] In other cases, the disruptive impact of technology, no matter how technically well designed, can be detrimental to the existing culture. Clearly, there are more considerations than just technical merit and efficiency that must be taken into account.

The book *When Helping Hurts* takes a hard look at international development efforts and includes warnings about the appropriate place and use of technology. One example given from the post–World War II era illustrates the concept of cultural appropriateness. At this time, several Western

---

[25]Monsma, *Responsible Technology*, p. 71.
[26]Egbert Schuurman, *The Technological World Picture and an Ethics of Responsibility* (Sioux Center, IA: Dordt College Press, 2005), p. 52.

experts concluded that peasant farmers in developing countries needed to adopt new crop varieties with higher yields. Although these crops had higher-than-average yields, they also had higher yield variations from year to year. Maximizing higher average yields makes sense on paper, but it can be disastrous if a single year of bad crops leads to starvation. In this context, it is more appropriate to choose a lower-risk crop, even if the average yield and efficiency of the crop may be lower.[27] The book argues against a "blueprint" approach to international development, instead emphasizing the importance of learning and of getting local people to participate in development projects.[28] This is necessary since local people know the most about their cultural context, including things that outsiders may not understand or appreciate. Appropriate technology will take into account cultural, environmental and social considerations in the design stage.

The norm of cultural appropriateness also involves weighing differing viewpoints of large scale and small scale, continuity and discontinuity, and centralization and decentralization.[29] Numerous issues arise when considering large-scale versus small-scale technology. In a book titled *Small Is Beautiful*, the author argues that bigger is not always better, and that small-scale technology can be a more appropriate choice.[30] On the other hand, complex software projects like desktop operating systems or the Internet are better suited as large-scale technologies. We need to recognize these tradeoffs as they arise and have the wisdom to discern what would be appropriate in a given situation.

The issue of continuity and discontinuity arises in software and hardware design. This includes considerations of backwards compatibility and compliance with standards. New software features and formats have advantages, but they must be weighed against the learning curve and retraining that might be required. Likewise, new computer paradigms can break away from the past, but they must be balanced with appropriateness. This includes introducing new computer technology in places where it was not

---

[27]Steve Corbett and Brian Fikkert, *When Helping Hurts* (Chicago: Moody Publishers, 2009), p. 116.

[28]Ibid., pp. 142-44.

[29]Monsma, *Responsible Technology*, p. 71.

[30]E. F. Schumacher, *Small Is Beautiful: Economics as if People Mattered* (New York: Harper & Row, 1973).

previously used. In each situation, we need to look not only at what might be gained but also at what might be lost.

One example is the introduction of new computer technology into a worship service. The use of new presentation technologies in church should be thoughtful, used with care and not distract from worship.[31] Use of multimedia technology should enhance participation rather than turning the congregation into an audience of spectators.[32] Technology should not be an intrusion to worship nor used for its own sake. Rather, it should serve the purpose of worship.

Various research studies are also investigating the appropriateness of screen-based computer technology in education. According to one study, screen-based technology improves "visual-spatial skills" and "visual intelligence" while at the same time weakening "mindful knowledge acquisition, inductive analysis, critical thinking, imagination and reflection."[33] It seems that certain types of knowing are enhanced, but at the expense of other ways of knowing. Therefore, students should have variation in their media diet, because each medium has advantages and disadvantages.[34] Schools need to reject the notion of a technological imperative in education and discern what computer technology does well while balancing it with other approaches to learning and instruction.

Finally, the issue of centralization versus decentralization is a relevant consideration in the age of the Internet and distributed computing. This is more than a technical choice; it includes the matter of where information resides and how it is controlled. For example, in cloud computing, data is migrated from local computers (decentralized) to large data centers in the "cloud" (centralized). Cloud computing has advantages in terms of backup, data sharing, and collaboration. But it has implications in terms of data security and reliable network access, and it requires trust that the privacy of the data will be respected. The issues of centralization versus decentralization are important considerations in an era of web computing.

---

[31]For more on technology and worship, see Quentin J. Schultze, *High-Tech Worship?* (Grand Rapids: Eerdmans, 2003).

[32]Brad J. Kallenberg, *God and Gadgets* (Eugene, OR: Cascade Books, 2011), pp. 116-17.

[33]Patricia M. Greenfield, "Technology and Informal Education: What Is Taught, What Is Learned," *Science* 323, no. 5910 (2009): 69-71.

[34]Ibid., p. 71.

In *Culture Making*, Andy Crouch suggests several questions to understand how an artifact fits into a culture. One of the questions is, "What does this cultural artifact make possible?" This is the primary question that is usually asked when evaluating new digital artifacts. However, another question is equally important, namely, "What does this cultural artifact make impossible (or at least very difficult)?"[35] Because computers make some things more difficult, this latter question ought to be asked as part of gauging the cultural impact of new computer technologies.

LINGUAL AND COMMUNICATION NORMS

The lingual norm involves issues of information, understanding and open communication.[36] Computer technology relies heavily on lingual information and communication systems. Open communication implies that there is a clear channel of information and dialogue among the parties involved. Open and clear information about product specifications and performance, as well as terms of use, can help customers make informed choices. This norm involves showing care for our neighbor and telling the truth when communicating.

Open and clear communication is critical for design teams. In his book *The Mythical Man-Month*, Frederick Brooks speaks about his experiences managing large technology projects at IBM, and a major theme is the importance of communication among teams. He suggests that teams meet often and in different ways, including informally, in structured meetings and by shared documentation in a workbook.[37] Even with multiple channels of communication, failures still occur. In a subsequent book, *The Design of Design*, Brooks observes that we cannot communicate perfectly because human beings are fallen.[38]

The linguistic norm also applies to user manuals, user interfaces and online help resources. The primary purpose of user manuals is to provide clear instructions, and the goal of a well-designed user interface is to help users understand clearly how a program operates. Error reports in software

---

[35]Andy Crouch, *Culture Making* (Downers Grove, IL: InterVarsity Press, 2008), pp. 29-30.
[36]Egbert Schuurman, *Faith and Hope in Technology: A Philosophical Challenge* (Toronto: Clements Publishing, 2003), p. 196.
[37]Frederick P. Brooks, *The Mythical Man-Month* (San Francisco: Wiley, 1995), p. 75.
[38]Frederick P. Brooks, *The Design of Design* (Boston: Addison-Wesley, 2010), p. 44.

systems should do more than a "core dump" by providing helpful and understandable messages and warnings.[39]

Communication norms are also relevant to programming languages, file formats and protocols.[40] For instance, standardized file formats can help people to communicate and exchange ideas more easily.[41] Considerations need to be made in programming languages so that programmers can use powerful yet unambiguous statements when coding. Early programming languages required programmers to write code using primitive low-level machine instructions. Many modern programming languages provide high-level instructions that enable programmers to write instructions more naturally.[42] Finally, to improve computer-to-computer communications, network protocols need to be concise and fault-tolerant so that computers can reliably exchange information. In large public networks like the World Wide Web, the protocols need to be open and well documented so different machines of different architectures from different vendors can enable people to electronically communicate with each other.

SOCIAL NORMS

The social norm is critical for users and developers of electronic communications to appreciate. This norm has to do with such things as courtesy, politeness and etiquette. This has implications for how people interact in person, as well as through electronic media. This norm ought to be shaped by the biblical norms of love and care.

This norm is an important one for those who interact with computer users and customers. Many jobs in the field of computing deal with customer service and sales. Social norms are important for those who work with customers and for whom courtesy and politeness are important attributes. In particular, help-desk and call-center staff members speak with users and walk them through technical troubles. People who contact help

---

[39]A core dump is usually a snapshot of the working memory of a computer program at the moment it has terminated abnormally. Although useful in debugging for software developers, it does nothing to inform a user about what went wrong.

[40]A protocol for digital messages is a set of rules for exchanging messages in or between computing systems.

[41]An example is the open document format (ODF), which was designed to provide an open standard for storing office suite documents.

[42]A good example is Python, a friendly, open source programming language.

desks are often in a flustered state, and courtesy and politeness can go a long way toward defusing a frustrating situation. Help-desk operators need to be patient with their less-knowledgeable callers, and callers need to show respect and patience as they are being assisted.

Social norms also come into play when programmers work together in teams. Interacting with coworkers and customers requires courtesy, politeness and patience. Even in the process of writing computer code, programmers should strive to produce hospitable code that is mindful of those who may need to read, maintain, modify or use the code in the future.[43]

The issue of politeness and etiquette includes the use of electronic communications. The lack of physicality or the feeling of anonymity associated with electronic messages may lead some to write things they would not say in person. Examples of ignoring social norms in online interactions include cyberbullying and flaming.[44] We should remember Paul's words to the Colossians: "Let your conversation be always full of grace, seasoned with salt, so that you may know how to answer everyone" (Colossians 4:6). Over time people have developed guidelines for online etiquette sometimes referred to as *netiquette*. Certain guidelines have been developed for online communications such as news groups, mailing lists, blogs and forums. Some of these guidelines have been described in a memorandum published by the Internet Engineering Task Force.[45] The use of mobile devices also requires a certain etiquette. Tethered to mobile devices in public places, people are distracted and absent from those who physically surround them.[46] Christians in an era of ubiquitous electronic communications need to recognize the importance of social norms for both online and face-to-face interactions.

Social networking sites have undergone remarkable growth since they emerged at the start of the millennium. The desire to connect and relate in community is a creational impulse, but doing so through an electronic filter changes things. Users and developers of social networking technologies

---

[43]Victor Norman, "Teaching How to Write Hospitable Computer Code," *Dynamic Link Journal* 3 (2011-2012): 10.

[44]*Flaming* is a common term for insulting exchanges between Internet users.

[45]See "Netiquette Guidelines," available online at www.rfc-editor.org/info/rfc1855.

[46]Sherry Turkle, *Alone Together: Why We Expect More from Technology and Less from Each Other* (New York: Basic Books, 2011), p. 155.

need to recognize that the lack of physicality inherent in electronic communications makes it a poor substitute for face-to-face interactions. According to Shane Hipps, "The digital space has the extraordinary ability to create vast superficial social networks, but it is ill-suited for generating intimate and meaningful human connection."[47] One of Marshall McLuhan's four "laws of media" states that when a medium is pushed to its limits, it tends to reverse its original characteristics. This is true also for social networking, which, when taken to an extreme, reverses into increased isolation rather than enhanced social interactions.

For reasons such as these, there are challenges when using the web for evangelism. In his book *God and Gadgets*, Brad Kallenberg recalls a story called *The Gospel Blimp* published by Joseph Bayly in 1960.[48] In this story, well-meaning Christians use a blimp equipped with loudspeakers and a large sign to broadcast the gospel to their town and to drop evangelistic tracts. George and Ethel pray that their unchurched neighbors will hear the gospel message through the blimp. But their neighbors actually find the blimp annoying. At the end of the story their neighbors become Christians, but not through the blimp. Rather it was through acts of kindness and through demonstrations of how the gospel was real to George and Ethel in their lives. Bayly concludes that impersonal means such as the blimp "are poor substitutes for personal communication of the gospel."[49] Although electronic media can be used to communicate the gospel, Kallenberg argues that "genuine communication requires *bodies*."[50]

New social norms are also being forged with developments in robotics. Sociable robotics "exploits the idea of a robotic body to move people to relate to machines as subjects."[51] Sociable robots are now being introduced as children's toys and for eldercare. Sherry Turkle, a social scientist who has explored these developments, discusses some of the issues of sociable robots in her book *Alone Together*. She observes how children quickly bond with sociable robots and the effects they have on people. She concludes, "Children need to be with other people to develop mutuality and empathy; interacting

---

[47]Shane Hipps, *Flickering Pixels* (Grand Rapids: Zondervan, 2009), p. 183.
[48]Kallenberg, *God and Gadgets*, p. 74.
[49]Joseph Bayly, *The Gospel Blimp* (Havertown, PA: Windward, 1960), p. 77.
[50]Kallenberg, *God and Gadgets*, p. 79.
[51]Turkle, *Alone Together*, p. 44.

with a robot cannot teach these."[52] Robots are also being introduced for eldercare to provide help with daily chores, dispensing medication and providing companionship. Some argue that robots "will be more patient with the cranky and forgetful elderly than a human being ever could."[53] But can a machine provide real care if it is incapable of caring? Some have explored the notion that it is deceptive to build robots with features that give the illusion that people can form relationships with them.[54] Machines are certainly no substitute for human interaction. The duty to show love and care to others, especially to more vulnerable people like children and the elderly, should not be offloaded to machines.

Economic Norms

The economic norm deals with stewardship and the wise use of resources. The notion of stewardship is also a clear biblical norm. This norm has implications for both users and developers of computer technology. The economic norm deals with stewardship of material and energy, as well as human resources.[55]

Computer technology raises many stewardship issues with regard to materials and waste handling. The cycle of constant upgrading of computers and cell phones and other electronic devices has generated a substantial amount of e-waste (electronic waste).[56] Many devices include toxic materials that require proper disposal and must be recycled in a responsible manner. Various electronic devices incorporate components with rare-earth materials that have a limited supply and hence need to be used wisely and recycled when the product reaches end-of-life. The vast marketing machine that drives consumer electronics creates "artificial needs" rather than using the earth's limited resources to meet "genuine needs." We are bombarded daily with messages coaxing us to purchase the latest computers and smartphones. This is exacerbated by planned obsolescence, whereby com-

---

[52]Ibid., p. 56.
[53]Ibid., p. 106.
[54]Amanda Sharkey and Noel Sharkey, "Children, the Elderly, and Interactive Robots," *Robotics and Automation Magazine, IEEE* 18, no. 1 (March 2011): 32-38.
[55]Schuurman, *Faith and Hope*, p. 197.
[56]Mark Anderson, "What an E-Waste," *IEEE Spectrum* 47, no. 9 (September 2010): 72.

puter products are designed with a limited life to increase sales.[57] Stewardship of resources should be an important consideration for both users and developers of computer technology.

Although God has created many good things for us to enjoy, we need to do more to reduce waste and to live in a more sustainable fashion. Individual Christians need to act locally but think globally. This begins with reducing and properly recycling our own e-waste while protesting practices such as shipping hazardous waste to developing countries. Support should be given to international agreements, such as the Basel Convention, which regulate exports of hazardous and other wastes and which promote sound environmental management.[58] In fact, even in the design stage engineers ought to consider how a product will eventually be disposed of and how material may be recovered or recycled. This involves performing a full lifecycle analysis as part of the design considerations and working toward greater sustainability. We should support "green computing" efforts to make computing more sustainable. This includes efforts to reduce power demands for large data centers and make personal computer equipment run more efficiently. It involves the design of lower-power electronics, as well as the design of programs and algorithms that can run more efficiently. Responsible computer technology will intentionally include considerations to improve sustainability.

Human resources are also important to steward properly. Computer and electronic manufacturing jobs, including many that have been "offshored" to developing countries, need to provide safe and fair conditions for workers. In some sectors, computers, robots and automation equipment have displaced a massive number of jobs. Some of these developments have been good: for instance, using robots to take on dangerous and highly repetitive jobs unsuited for humans. Robots can also work in hazardous environments, such as mines and in space, thus reducing the risk for humans. Although many developments in computer technology have produced new jobs and opportunities, many people and communities have been left behind. When contemplating new technologies, we should be concerned not only about efficiency but also the impact on human flourishing.

---

[57]Kallenberg, *God and Gadgets*, p. 111.
[58]See www.basel.int for more information about the Basel Convention.

When the economic aspect is absolutized, maximum profit is gained "at the expense of the individual person and at the cost of overlooking the dangers to the environment posed by waste products."[59] This is not to say that profits are wrong, but they should be made "in connection with service to God and service to one's neighbor."[60]

AESTHETIC NORMS

The aesthetic norm deals with the notion of delightful harmony.[61] Good technology is characterized by being a joy and delight to use. Nicholas Wolterstorff considers the example of a spade and suggests that a spade serves its purposes well if it consists of two things.[62] First, it must be good for digging holes, but second, it should be good and satisfying to use for this purpose. The fusion of function and beauty captures the meaning of delightful harmony.[63] In *The Design of Everyday Things*, Donald Norman suggests, "Good designs will have it all—aesthetic pleasure, art, creativity—and at the same time be usable, workable, and enjoyable."[64] Likewise, computer products should be pleasing and intuitive to use. Another way of saying this is computer technology should be user-friendly, avoiding unnecessary complexity.

Aesthetic issues arise naturally in areas such as graphical user interfaces and dashboard designs. Often industrial designers or "user experience" professionals are employed to ensure that interfaces and devices are constructed in an attractive fashion. The form and function of a device should be in harmony, such that the form implies the function. Donald Norman concludes that "the appearance of the device must provide the critical clues required for its proper operation."[65] A good example is a computer mouse, which has a simple form but one that implies its function. This principle becomes more difficult to realize as more features are packed into small devices, a tendency sometimes referred to as "creeping featurism."[66] The

---

[59] Schuurman, *Technology and the Future*, p. 363.
[60] Ibid.
[61] Monsma, *Responsible Technology*, p, 73.
[62] Nicholas Wolterstorff, *Art in Action: Toward a Christian Aesthetic* (Grand Rapids: Eerdmans, 1980), p. 156.
[63] Monsma, *Responsible Technology*, p. 74.
[64] Norman, *Design of Everyday Things*, p. xiv.
[65] Ibid., p. x.
[66] Ibid., pp. 172-74.

addition of more features can increase complexity and dramatically reduce usability. Norman describes this issue in *The Design of Everyday Things*: "It is true that as the number of options and capabilities of any device increases, so too must the number and complexity of the controls. But the principles of good design can make complexity manageable."[67]

A classic example of a poor and anti-intuitive interface from the past is that of videocassette recorders (VCRs).[68] Some older people may recall that many VCRs were often left displaying a flashing 12:00 instead of the proper time, because it was notoriously difficult to set the date and time. To help make a memorable point to his students, computer science professor Randy Pausch would bring a sledgehammer to class to destroy a VCR.[69] Technology should not only be functional but should be designed for ease of use.

The whole area of human-computer interaction (HCI) explores issues that occur at the user interface. The user interface should not be distracting but should provide a helpful layout of controls that are intuitive to use. The area of ergonomics is dedicated to design with attention to optimizing human well-being in addition to overall system performance. A good design will reduce physical strain and will consider the well-being of the operator. A good interface design provides the user with controls that are both intuitive and elegant. A pleasing font design should provide fonts that are both attractive and easy to read. Website designers also need to be aware of how layout, color, menus and style affect the attractiveness and ease of navigation of a website. In the words of Egbert Schuurman, "Technology ought always be humanity's servant. Thus human beings ought not to have to adapt themselves to computer systems, but vice versa."[70]

Even the underlying computer code and computer architectures have aesthetic aspects and style. Frederick Brooks talks about the role of style in technical designs and how people can recognize the distinctive style of a Seymour Cray computer or how programmers can be identified by the style of their code.[71] Many Apple computers and products have illustrated a de-

---

[67]Ibid., p. 31.
[68]For younger readers who may not be familiar with the VCR: the VCR was a device used to record analog video and audio signals on a magnetic tape.
[69]Randy Pausch, *The Last Lecture* (New York: Hyperion, 2008), p. 150.
[70]Schuurman, *Faith and Hope*, p. 198.
[71]Brooks, *The Design of Design*, p. 147.

lightful attention to aesthetics and usability. One Apple computer, the distinctive Power Mac G4 Cube, ended up on display in the New York Museum of Modern Art.[72]

A clear and consistent style is not only more elegant, but it can make designs easier to understand and maintain. Messy *spaghetti code* and poorly laid-out schematic diagrams may not be in error, but they are difficult to read, comprehend and verify.[73] Some programming languages include a GOTO statement that allows control to jump or branch to other parts of the program. If used inappropriately, program flow becomes hard to follow as control hopscotches throughout a program. In a letter to the editor of the *Communications of the ACM* under the title "Go To Statement Considered Harmful," Edsger Dijkstra criticized the use of the statement and advocated structured programming.[74] Dijkstra contends that this particular programming statement "is just too primitive; it is too much an invitation to make a mess of one's program."[75] Programming languages should not only facilitate the writing of code that is correct but also code that is elegant. Electronic designs can also be messy. An untidy rat's nest of electronic wires may in fact be correct, but it is not pleasing to view or debug. The delightful harmony of beautiful form and function can produce programs and circuits that not only look better but that are easier to maintain and repair.

In his 1974 Turing Award lecture, entitled "Computer Programming as an Art," the respected computer scientist Donald Knuth stated that "the chief goal of my work as an educator and author is to help people learn how to write beautiful programs."[76] Indeed, there can be delight in writing a beautiful program or designing an elegant electronic circuit. Like other modalities, the aesthetic modality is at play when we work with computer technology.

---

[72]See "The Collection," Museum of Modern Art website, www.moma.org/collection/browse_results.php?object_id=82134.

[73]*Spaghetti code* refers to computer source code that is not well structured but is complex and tangled.

[74]Edsger W. Dijkstra, "Go To Statement Considered Harmful," *Communications of the ACM* 11, no. 3 (March 1968): 147-48.

[75]Ibid., p. 147.

[76]The Turing Award is an annual award given by the Association for Computing Machinery (ACM) to an individual selected for making technical contributions to the computing community. It is widely recognized as the highest distinction in computer science. See Donald E. Knuth, "Computer Programming as an Art," *Communications of the ACM* 17, no. 12 (December 1974): 670.

JURIDICAL NORMS

The juridical norm deals with issues of justice, which is a fundamental biblical norm. To act justly is one of the things that the Lord requires of us (see Micah 6:8). Developers of computer technology need to carefully consider whether their products enhance justice or whether they contribute to forms of injustice. This includes justice to plants, animals, raw materials and also to people and developing nations.[77] Governments also have a role to create legislation to help guide technological developments with respect to areas such as compliance with safety, electrical standards and environmental protection, to name just a few.[78]

Computer technology figures prominently in discussions about the so-called digital divide, which refers to the lack of access to computers between different groups. The digital divide in developing countries is just one manifestation of the wider issue of justice and poverty that includes more basic needs, such as safe water, nutrition and sanitation.[79] Often those with technology are privileged, while others are shut out or left behind. Projects like the WiderNet Project provide resources, training and computers to schools, clinics and libraries in places such as parts of Africa with poor digital communication resources. An example of one of their projects is the eGranary Digital Library, a computer system that provides access to millions of digital documents without requiring a connection to the Internet. The Scriptures are clear about the concern we ought to have regarding justice for the poor and less fortunate (see Proverbs 29:7). Christians ought to support efforts that provide culturally appropriate and sustainable computer technology in developing countries.[80]

The lack of access and opportunities is not necessarily confined to developing countries. Author Jane Margolis has explored the lower numbers of

---

[77]Schuurman, *Faith and Hope*, p. 198.

[78]Schuurman, *Technological World Picture*, p. 60.

[79]Kumar Venkat, "Delving into the Digital Divide," *IEEE Spectrum* 39, no. 2 (February 2002): 14-16.

[80]For information about the WiderNet Project, see www.widernet.org. I saw firsthand how helpful the eGranary Digital Library could be when I traveled to assist a fledgling Christian university in West Africa. In a context where the library holdings were limited and dated and Internet access was difficult, this digital library provided an important resource for students. See www.widernet.org/eGranary. For some examples of projects that demonstrate design innovations for poorer nations, see Cynthia E. Smith, *Design for the Other 90%* (New York: Cooper-Hewitt National Design Museum, 2007).

racial minorities in computer science in the United States in *Stuck in the Shallow End: Education, Race, and Computing*.[81] This book looks at the experiences of students and teachers in several public high schools and the differences in educational opportunities. In an economy that is increasingly knowledge-based, access to quality computer education should not be limited to the rich and privileged.

It is not just the lack of technology that is a source of injustice; the applications of technology can also be unjust. The growing use of digital tracking and surveillance technology can be invasive and intrusive. These days many of our activities are monitored through the data trails that we leave behind. Databases record our financial transactions, our employment history and our health records. Websites use *cookies* to track our online activities.[82] Smartphones equipped with GPS (Global Positioning System) features are able to track and report our physical locations. RFID (Radio Frequency Identification) tags are growing in popularity and are currently being used in various applications ranging from passports to book identification in libraries.[83] Many predict that with the continuing growth of RFID tags, there will be many new threats to personal information and privacy. Everywhere, it seems, our data is being mined. But is the right to privacy really a justice issue? Should Christians be concerned about their privacy?

We are called to "give back to Caesar what is Caesar's," and indeed this may very well include some of our information (Matthew 22:21). A government needs certain kinds of information in order to govern effectively and to make good decisions. In fact, statistics, polls and opinion surveys are often used to shape government policies. Some of the information in our personal sphere, however, falls outside the legitimate sphere of government. In some cases, there may be less intrusive means for governments to gain the information they need. Concerns over crime and terrorism have brought new tensions between privacy and security. The collection of sensitive information may be seen as overreaching and can sometimes lead to abuse.[84]

---

[81]Jane Margolis, *Stuck in the Shallow End: Education, Race, and Computing* (Cambridge, MA: MIT Press, 2010).

[82]A *cookie* is a piece of data a website stores on a visitor's computer and can be used to track a user's activities.

[83]RFID tags allow for the tracking of nearby objects using radio waves.

[84]For some examples, see Sara Baase, *A Gift of Fire: Social, Legal, and Ethical Issues for Comput-*

Governments need to carefully weigh privacy interests on one hand and the legitimate need for information on the other.

The same is true for information collected by commercial companies. In order to do business with a company, a legitimate exchange of information needs to take place. For instance, ordering a product from a company may require you to provide your name and a delivery address. However, attempts to collect certain additional personal information beyond this may not be justified, especially if it leads to stalking a customer's online activities and the nuisance of ongoing unsolicited marketing activities, such as junk email. Selling or using personal information for purposes other than those that were intended is a breach of trust. Companies need to observe privacy practices that respect the visitors and customers who use their website. When companies legitimately collect data, it should be stored securely and kept from misuse. Furthermore, websites ought to use clear language and be upfront about their privacy policies.

Juridical norms also figure prominently in issues surrounding intellectual property. Intellectual property includes legal concepts such as copyrights, patents, trademarks and trade secrets.[85] Copyrights and patents figure prominently in the field of computing. Intellectual property deals with intangible works or ideas rather than physical objects. The buyer who purchases a song, movie or a piece of software is only paying for a license to use it. This has become a thorny issue in the age of computer networking, when perfect digital copies can be made easily and widely distributed in high volumes over the Internet. This is further complicated by the fact that the Internet spans international boundaries. People make numerous excuses for illegal digital copying, including arguments such as not being able to afford it, that everyone else is doing it, or that the company selling the product is wealthy enough already.[86] None of these excuses is valid. Although digital copies can be downloaded and copied easily, intellectual property still has ownership. Intellectual property laws are intended to protect creators of content such as music, books, software and movies so that they can receive compensation for their work. Although intellectual property is different than physical property,

---

ing Technology, 4th ed. (Upper Saddle River, NJ: Prentice Hall, 2013), pp. 84-95.
[85]Baase, *A Gift of Fire*, p. 180.
[86]Ibid., p. 189.

the command "thou shall not steal" still applies. Digital content is protected by law, and Christians need to respect the governing authorities when it comes to intellectual property laws (see Romans 13:1).

One form of intellectual property is a *patent*. Simply put, a patent grants an inventor of a machine or process exclusive rights for a fixed period of time to build that invention. The purpose of patents is to reward innovators who invest in research that leads to useful products. Justice in patent law means considering what things are patentable and what things ought not to be patented. For instance, is it just to patent life forms or mathematical equations, which are not really inventions but discoveries about existing aspects of creation? Like other written works, computer programs are protected by copyrights, but applying patents to software raises many questions. Should computer algorithms, which are essentially mathematical operations, be patented and owned?[87] Because of software patents, unsuspecting programmers can be threatened with litigation for writing code that inadvertently violates questionable patents.[88] Patent laws need to be just, upholding the rights of the inventor and encouraging innovation but without hindering the legitimate rights of others.

Another form of intellectual property protection is the *copyright*. Simply put, a copyright is granted to an author of literary, musical, artistic or other written work (such as a computer program) that allows them the exclusive right to reproduce and distribute the copyrighted work. Copyright protection has led to the controversial rise of DRM (Digital Rights Management) technology to control access to digital content and devices. Although DRM protects content providers, it remains controversial because of its inconvenience for legitimate users.[89] Copyright laws need to properly protect authors while at the same time preventing nuisance litigation against legitimately licensed users. For instance, legitimate users should be allowed to make personal backup copies or convert from one digital media format to another without fear of violating copyrights. It is also important to maintain the notion of fair use to ensure limited use of copyrighted material

---

[87]Ben Klemens, *Math You Can't Use: Patents, Copyright, and Software* (Washington, DC: Brookings Institution Press, 2005), pp. 4, 26.

[88]Ibid., pp. 87-91.

[89]For this reason, some have suggested that DRM should be described as Digital *Restrictions* Management.

for commentary, criticism, research or teaching. Likewise, computer users have a responsibility to respect copyrights and ownership of digital content.

An interesting development has been the growth of free and open source software, which provides an alternative to the traditional commercial or proprietary software model. Proprietary software is normally sold without the *source code*, which is the language in which the program is actually written. In this way, companies keep the inner workings of their program secret to protect their product and prevent others from duplicating it. This has been compared to buying a car with the hood welded shut.[90] Although this helps protect the intellectual property of the company, many argue that this is a disadvantage for end users who are dependent on the company's effective monopoly on any software fixes, updates or enhancements.

Proprietary software typically comes with a restrictive EULA (End-User Licensing Agreement), which few people bother to read. These EULAs usually include clauses forbidding the software from being copied or reverse-engineered. They also typically include broad limitation of liability statements, which basically remove any responsibilities the company may have to ensure the software works in terms of its merchantability or fitness for a given purpose. Unlike other products, software rarely comes with guarantees but rather with disclaimers. Critics point out that proprietary software protected by a restrictive EULA limits the rights of users while simultaneously limiting the responsibilities of the company.

In recent years, a vast amount of high-quality free software has been developed, ranging from free operating systems such as Linux and office suites such as LibreOffice to software that runs on devices like smartphones. Free software is an idea supported by a loosely knit network of programmers and companies from around the world who give their code away for free. Because the source code is available and open, this movement has also been called the *open source movement*. In this paradigm, the intellectual property rights are set aside for the advantage of copying, sharing and collaborating. In *The Cathedral and the Bazaar*, Eric Raymond describes the open source development model as a bazaar, in which the software development process occurs in view of the public. The proprietary software model he likens to a cathedral,

---

[90]See Bob Young, "Open Source Is Here to Stay," *ZDNet*, May 3, 2000, www.zdnet.com/news/open-source-is-here-to-stay/107280.

which is characterized by a tightly controlled process in which access to source code is restricted to an exclusive group of software developers.[91]

The free software movement has turned the notion of copyright upside down by defining something called *copyleft*. Copyleft uses copyright law in the opposite way it is normally used: namely, to protect freedoms in software. Under copyleft, the developer copyrights the program and freely releases it under a licensing agreement that encourages others to use or modify the software with the requirement that all modified versions of the program must remain free as well.[92] An example of copyleft is the GNU General Public License (GPL), a popular free software license originally written by free software activist and programmer Richard Stallman. The philosophy of the GPL involves four basic freedoms. These freedoms are as follows:[93]

1. the freedom to run the program, for any purpose

2. the freedom to study how the program works, and change it so it does your computing as you wish

3. the freedom to redistribute copies so you can help your neighbor

4. the freedom to distribute copies of your modified versions to others

The term *free software* refers to these freedoms rather than cost. In fact, the free software definition includes explicit provision to allow the software to be sold or to charge for its distribution, but the freedom to copy and change the software must be maintained. The free software definition puts it this way: "you should think of 'free' as in 'free speech,' not as in 'free beer.'"[94] The idea of copyleft has extended to other published works such as documents, music and images.[95]

The open source paradigm has many aspects with which Christians can sympathize. The notion of helping your neighbor is a wonderful concept reflected in sharing software that can be duplicated and modified with ease. This

---

[91]Eric S. Raymond, *The Cathedral and the Bazaar: Musings on Linux and Open Source by an Accidental Revolutionary* (Sebastopol, CA: O'Reilly, 2001).

[92]The GNU GPL is an example of a copyright document that implements copyleft. See www .gnu.org/licenses for more information.

[93]The freedoms can be found in the GNU free software definition at www.gnu.org/philosophy/ free-sw.html.

[94]"What Is Free Software?" GNU Operating System, www.gnu.org/philosophy/free-sw.html.

[95]Many of these works can be found online at sites such as the Creative Commons, http:// creativecommons.org.

can be of particular benefit to those on the other side of the digital divide, such as those in developing countries. The notion of common grace is evident in the delightful software that is created and shared by a diverse community of people collaborating around the world. Furthermore, Christians can actively partic-ipate in open source projects, giving shape to software that can benefit our neighbors by sharing useful tools that open up their workspace.

While there are those who contribute code as a hobby, it is also proper that open source software developers can earn a living. Proprietary software companies can include the costs of paying their programmers in the sticker price of their products. The means to pay programmers in open source projects is not always clear, but some approaches have emerged. Funding for open source efforts can be achieved through creative business models such as the "software as a service" approach, in which customers pay for the service and hosting rather than for the software itself. The GPL specifically includes a clause permitting fees for distributing free software. Other funding sources include open source software foundations, donations, advertising and contributions from companies that benefit from open source software.

Both proprietary and open source software have a place. Frederick Brooks acknowledges the virtuous aspects of open source software but notes that there is still a need for carefully architected, controlled and tested designs, such as is the case for national air-traffic control systems.[96] It is also true that many successful open source projects (like Linux) have benefited from ideas originally developed in proprietary software. Likewise, proprietary software (such as web browsers) has adopted features originally developed in open source projects. The cathedral and the bazaar are both legitimate and have their respective places in the software landscape.

ETHICAL NORMS

The moral norm deals with showing love and care for our neighbor.[97] The call to love our neighbor is a key biblical norm. This has implications for those who produce technology and software systems because computer technology is value-laden, and it includes moral and ethical dimensions.

The use of computer and web technology to distribute pornography is

---

[96]Brooks, *The Design of Design*, pp. 55, 57.
[97]Monsma, *Responsible Technology*, p. 74.

a common example of a moral issue. In this case, creational aspects including technology and sexuality are misdirected and distorted. Gambling is another example of an activity that has become more accessible through the web. The use of content-filtering technology can be a useful tool to help reduce access to inappropriate material, but this is not just a technical issue that can be solved using a technical solution. Both pornography and gambling can lead to addiction and are destructive influences in people's lives.

Moral norms are not limited to identifying certain activities as immoral; such norms can also guide the positive use of technology to show love and care to others. Frederick Brooks makes a direct connection between his faith and what he does as a computer scientist: "There are certain common ethical standards in science, but Christianity makes you more concerned about carrying them out. It guides you to work on things you think are important."[98] Christian developers should choose work based on a desire to provide products that ethically serve needs in our society. This norm is particularly important to keep in mind when the technological imperative pressures us to do things because we *can*, rather than because we *ought* to do them.

Computer technology can be designed with care for people in mind. For example, the design of electronic devices should take into account ergonomic considerations such as displays that reduce eye strain. Software development that takes into account accessibility features for the visually impaired is an example of showing care for persons. In fact, computer technology can assist disabled persons with a variety of tools, such as advanced wheelchairs and text-to-speech and speech-recognition systems.

Care for persons implies that there are some things that we should not delegate to computers. Joseph Weizenbaum, the creator of the ELIZA program that simulates a Rogerian psychotherapist, was alarmed at suggestions that a computer could one day be used to automate psychotherapy. Weizenbaum observes, "I had thought it essential, as a prerequisite to the very possibility that one person might help another learn to cope with his emotional problems, that the helper himself participate in the other's experience of those problems, and, in large part by way of his own empathetic

---

[98]Dennis Shasha and Cathy Lazere, *Out of Their Minds: The Lives and Discoveries of 15 Great Computer Scientists* (New York: Copernicus, 1998), p. 174.

recognition of them, himself come to understand them."[99] In a similar way, a child-minding robot cannot replace human interaction, just as a television makes a poor babysitter. Weizenbaum concludes that "there are limits to what computers ought to be put to do."[100] The norm of love and care will mean that there are some software projects that simply ought not to be pursued—ever.

The sixth commandment, "You shall not murder," speaks to the responsibility of programmers for safe and reliable products, especially when they control systems upon which human life and safety rely (Deuteronomy 5:17). We must not only ensure that our work will not harm people, but in our work we need to actively live out the command to love our neighbor as ourselves. This involves employing proper design practices and using care when designing products. It may also mean refusing to do work that one is not competent to perform properly. The often-cited Therac-25 incidents are an example in which poor software design practices contributed to the serious harm and death of several people.[101] The Therac-25 was a radiation therapy machine used to treat people with cancer that was produced by Atomic Energy of Canada Limited. Between 1985 and 1987, this machine was involved in several accidents that resulted in six patients receiving massive overdoses of radiation. The accidents resulted in severe and painful injuries, and some patients later died as a result. Subsequent studies uncovered numerous factors that contributed to the accidents. Some of these contributing factors included poor management, lack of procedures, an overconfidence in the software system, poor software engineering practices and inadequate safety interlocks.[102] This case illustrates the potential consequences of poor work in safety-critical systems. A situation such as this should serve as a stark reminder of the ethical responsibilities of technicians, programmers, engineers and managers in their work.

Some organizations, such as the Association of Computing Machinery (ACM) and the Institute of Electrical and Electronic Engineers (IEEE),

[99]Joseph Weizenbaum, *Computer Power and Human Reason: From Judgment to Calculation* (New York: W. H. Freeman, 1976), pp. 5-6.
[100]Ibid., p. 11.
[101]Baase, *A Gift of Fire*, pp. 377-83.
[102]Nancy G. Leveson and Clark S. Turner, "An Investigation of the Therac-25 Accidents," *Computer* 26, no. 7 (1993): 38.

have published codes of ethics for professionals.[103] Many of these ethical principles are ones that Christians also embrace. For instance, these ethical codes include statements affirming honesty in making claims or estimates. This principle is also stated in Proverbs, when the writer says that "the LORD detests dishonest scales" (Proverbs 11:1). The ethical codes also deal with avoiding injury to others and safeguarding human health and the environment. They address the need to understand technology and its consequences (a statement that implicitly acknowledges that technology is not neutral) and to ensure competency in work that is performed. Typically, licensed professional engineers must comply with a code of ethics set out by statutes or by professional regulatory bodies. Such duties are sometimes referred to as *fiduciary responsibilities*, in which engineers must exercise a standard of care to ensure the well-being of the public. These codes often cover areas such as ensuring public welfare, whistle-blowing, avoiding conflicts of interest and working only in areas in which one is competent. Christians ought to encourage all such efforts that are consistent with the call to show love and care for our neighbor.

Social and ethical norms also figure prominently in the area of video games, especially when many games feature increasing levels of graphic violence. "First-person shooter" video game genres, which have become extremely popular, enable a player to vicariously experience violent actions through the eyes of a game character. Such games are often rated "M" for "Mature"—an ironic term for games that glamorize violence or combat. Many young boys and teens are powerfully drawn to these types of games, and some play them compulsively. Over the years, various studies have reported evidence of adverse effects from these games, such as an increase in aggressive behavior. How should one's faith inform one's use of computer games? What does it mean to be a Christian game developer? If violence is part of life, does this justify its inclusion in our games? What would a well-designed video game look like if the developers constructed it to reflect the various norms?

Some films portray violence as part of a story, illustrating the reality of evil and the consequences of wrongdoing in a broken world. But the experience changes in a video game. The first-person construct creates a com-

---

[103]The IEEE code of ethics can be found at www.ieee.org/about/ethics_code. The ACM code of ethics is available at www.acm.org/about/code-of-ethics.

pletely different context in which the player actively participates in the simulated violence. In Philippians, we read the following directive: "Finally, brothers and sisters, whatever is true, whatever is noble, whatever is right, whatever is pure, whatever is lovely, whatever is admirable—if anything is excellent or praiseworthy—think about such things" (Philippians 4:8). We need to think more about what is "excellent," "praiseworthy" and "true" in our video games and other electronic entertainment. Christian game developers can help by using their gifts to expand the space and paradigm that video games encompass and to explore normative ways to entertain video game players. Aesthetic philosopher Calvin Seerveld suggests "fashioning games that are not cruelly competitive, but full of surprise for catching a child with discoveries unawares."[104]

Other moral issues arise when computer technology is deployed on the battlefield. Tele-operation of robots is now commonplace in the military, with unmanned aerial and land vehicles controlled by a remote human operator.[105] Tele-operated war planes and vehicles can attack a target when a person pulls the trigger by remote control. But this technology has repercussions. The technology that enables soldiers to operate over great distances also tends to reduce remote combatants to targets on a screen.[106]

The situation becomes even more complex as robot drones become more autonomous. As such systems become more technically feasible, autonomous robots taking lethal action on their own becomes a possibility.[107] Should a robot be permitted to pull the trigger without a human in the loop? If autonomous robots remove human soldiers from the battlefield, will the perception of a "risk-free" war make technically advanced nations less hesitant to enter combat? Should the decision to take a human life on the battlefield be relegated to a machine? Decisions such as these are major moral decisions, and even in combat, soldiers bear

---

[104]Calvin G. Seerveld, *A Turnabout in Aesthetics to Understanding* (Toronto: Institute for Christian Studies Publication, 1974), p. 21.

[105]P. W. Singer, *Wired for War: The Robotics Revolution and Conflict in the 21st Century* (New York: Penguin, 2009).

[106]Kallenberg, *God and Gadgets*, p. 108.

[107]Lora G. Weiss, "Autonomous Robots in the Fog of War," *IEEE Spectrum* 48, no. 8 (August 2011): 31-34, 56-57.

individual responsibility for their actions. In fact, some have argued that a condition for fighting a just war is that someone can be held justly responsible for the deaths that occur.[108] In *Governing Lethal Behavior in Autonomous Robots*, Ronald Arkin explores implanting responsible ethical decision making in autonomous lethal robots and proceeds to describe the architecture for an "ethical governor" for lethal robots.[109] The desire to safeguard noncombatants is commendable, and work such as this does highlight the need to take ethical issues seriously. However, the technology also raises many serious questions. What are the rules of engagement for unmanned lethal machines? Who is responsible for the actions of a lethal robot soldier? Can robots commit war crimes? Can ethical behavior be distilled into a step-by-step set of rules? Algorithms exist for playing chess and for computing square roots, but can there be an algorithm for ethical behavior? People disagree about ethics and how it applies to different situations, but assuming a consensus could be found on a set of algorithmic principles, would this ever be sufficient?

In his science fiction novels, Isaac Asimov introduces three fundamental "Rules of Robotics."[110] They are as follows:

1. A robot may not injure a human being or, through inaction, allow a human being to come to harm.

2. A robot must obey any orders given to it by human beings, except where such orders would conflict with the First Law.

3. A robot must protect its own existence as long as such protection does not conflict with the First or Second Laws.

Are these rules a sufficient algorithm for ethical behavior? Some people believe that the human mind is, in fact, a computer of sorts; for them, the notion that a computer could be constructed to make ethical decisions is not a far leap. If that were the case, a computer could make ethical decisions faster, more efficiently and with less bias than a human. Isaac Asimov's stories include several plots that hinge on the unexpected

---

[108]Noel Sharkey, "Automated Killers and the Computing Profession," *Computer* 40, no. 11 (2007): 122.

[109]Ronald C. Arkin, *Governing Lethal Behavior in Autonomous Robots* (Boca Raton, FL: Chapman & Hall/CRC, 2009), pp. 130-33.

[110]Isaac Asimov, *I, Robot* (New York: Bantam Dell, 2004), pp. 44-45.

results that occur when robots encounter situations in which these laws are in conflict.

Autonomous robots on the battlefield would require far more than exceptional sensors and the ability to classify objects. Such a robot will have to "make decisions in complex and entirely unpredictable circumstances."[111] The modal aspects introduced in chapter 2 are a useful tool to remind us that moral norms cannot be reduced to a numeric modality such as an algorithm. Ethical thought requires more than flow charts and algorithms. The notion of allowing a robot to make moral decisions on behalf of a human, especially ones that deal with life and death decisions, reveals many ethical and philosophical presuppositions. Computer scientist Noel Sharkey puts it bluntly: "Ultimately, we must ask if we are ready to leave life-or-death decisions to robots too dim to be called stupid."[112] We need to appreciate the complexity of ethical norms and call into question any attempts to relegate moral decisions to a machine.

The whole field of military technology raises many difficult ethical questions. A few robot researchers have even gone so far as to avoid accepting research funds from military sources.[113] But even if robot research is not funded by the military, there is nothing to prevent published research from being used for military applications. Bill Joy, the former chief scientist at Sun Microsystems, has suggested that the only alternative is relinquishment of certain kinds of research: that is, "to limit development of the technologies that are too dangerous, by limiting our pursuit of certain kinds of knowledge."[114] This is an area about which even Christians disagree. Some Christian traditions practice pacifism, while others suggest that loving our neighbor and bringing justice requires action by the state, which has been granted the "sword" (see Romans 13:4). Many Christians recognize the notion of a just war and the necessity of fighting to defend oneself or to defend the weak and oppressed. One thing is clear: in a perfect world before the fall, technology was not intended to build weapons. As Christians, we long for the day when Christ returns and na-

---

[111]Sharkey, "Automated Killers," p. 122.
[112]Ibid., p. 123.
[113]Singer, *Wired for War*, pp. 171-73.
[114]Bill Joy, "Why the Future Doesn't Need Us," *Wired Magazine* 8, no. 4 (April 2000): 238-64.

tions "will beat their swords into plowshares and their spears into pruning hooks" (Isaiah 2:4). Even if one accepts the notion of a just war, societies need to discern when there may be too many resources directed toward products for war. With considerable technical resources being directed toward "swords," perhaps more efforts should be directed to developing "plowshares" and other peaceful products. The problem becomes amplified when people begin to put their trust in military strength and technological superiority, which are also idols of our time.[115]

FAITH NORMS

The final norm is the faith norm, which is related to trust. This norm has two aspects.[116] The first is that developers of computer technology can work in the confidence that the creational laws upon which computers depend will continue to hold. Furthermore, users should be able to have confidence that computer products are designed properly for their intended use and that they can be used safely. If a computer device advertises certain specifications, the claims about features and performance should be dependable and not misleading.

The second aspect of this norm is more significant and deals with where we place our ultimate trust. The faith norm has a definite direction—faith can be directed to God or elsewhere. In essence, this is a matter of the heart. We should, in the words of Egbert Schuurman, be "led by the belief that humanity is called to the task of technology and that people are obliged to accept this mission as a responsibility before God." Schuurman writes, "The meaning of technology is service to God."[117] In Colossians 1 we read that all things in heaven and on earth were not only created through Christ but that they were also created for Christ (see Colossians 1:16). In the case of technicism, ultimate trust is placed in technology and its meaning and purpose are perverted and distorted.[118] A faith in technology leads to all kinds of negative consequences. We are created to be people of faith, and our faith has a

---

[115]Bob Goudzwaard, *Idols of Our Time* (Downers Grove, IL: InterVarsity Press, 1984), pp. 106-7, 61-77.

[116]Monsma, *Responsible Technology*, p. 75.

[117]Egbert Schuurman, *Technology and the Future: A Philosophical Challenge* (Toronto: Wedge Publishing Foundation, 1980), pp. 365, 361.

[118]Ibid., pp. 359-61.

direction. Augustine stated that our hearts remain restless until they rest in God. Despite all the technological wonders that surround us, this remains the case today.

CONCLUSION

We began this chapter with Colossians 1, which clearly states that Christ is central in creation. Lesslie Newbigin writes, "Jesus is the clue for understanding all that is."[119] It was through him and for him that all things were created, and it is in him that all things hold together. It is through Christ's redemptive work on the cross that all things in creation are being reconciled, and this has implications for all areas of life, including computing (see Colossians 1:20). Albert Wolters writes, "Christ is the reconciler of all things, and if we have been entrusted with 'the ministry of reconciliation' on his behalf (2 Corinthians 5:18), then we have a redemptive task wherever our vocation places us in his world."[120] This includes the area of computer technology.

Responsible computer technology involves following Christ and discerning God's purpose for creation. We must strive to be more like Christ in all areas of life, including our technical activities. Because computer technology is value-laden, Christians need to be attuned to the values and norms that are at play in the design and use of computer technology. The normative modal aspects include the historical (cultural), lingual, social, economic, aesthetic, juridical, ethical and faith aspects. There are normative principles that apply to each of these areas that must be simultaneously pursued. Informed by a biblical worldview, Christians must strive to use and shape computer technology in responsible and normative directions. These activities need to be guided by the biblical norms of justice, stewardship, love and care, and with the overall goal of shalom. The following quote from *Our World Belongs to God* summarizes this well.

> Grateful for advances
> in science and technology,
> we make careful use of their products,

---

[119]Lesslie Newbigin, *The Light Has Come: An Exposition of the Fourth Gospel* (Grand Rapids: Eerdmans, 1982), p. 3.
[120]Albert M. Wolters, *Creation Regained* (Grand Rapids: Eerdmans, 1985), p. 60.

on guard against idolatry
and harmful research,
and careful to use them in ways that answer
to God's demands
to love our neighbor
and to care for the earth and its creatures.[121]

---

[121]Article 52, *Our World Belongs to God: A Contemporary Testimony* (Grand Rapids: CRC Publications, 1988), p. 19.

5

Computer Technology
and the Future

It's tough to make predictions, especially about the future.

YOGI BERRA

■■■

The relatively short history of computer technology has brought numerous changes that were difficult to predict. Some famous predictions from the past now appear humorous, such as the one from *Popular Mechanics* in 1949 that made the bold prediction that "Computers in the future may . . . weigh only 1.5 tons."[1] In 1943 the chair of IBM, Thomas J. Watson, famously predicted, "I think there is a world market for maybe five computers."[2] Such predictions illustrate how wrong even the most technically astute people can be.

More recently, it was difficult to foresee the emergence of the World Wide Web and the widespread changes it has brought. Even those on the vanguard of new technology have trouble seeing the road ahead. In an *IEEE Spectrum* column, Robert Lucky compares technology to a bus: "The bus is being driven by someone up front whom we can't quite see, and worse yet, we're not even sure where the destination is supposed to be."[3] Where is this "bus" headed, and what part will technology play in our future world?

For those who do venture to speculate on the future of technology, the predictions are often characterized by either a sense of optimism or a sense of unease and even despair. Some see the world as advancing and getting

---

[1]Andrew Hamilton, "Brains That Click," *Popular Mechanics*, March 1949, p. 258.
[2]Sara Baase, *A Gift of Fire: Social, Legal, and Ethical Issues for Computing Technology*, 4th ed. (Upper Saddle River, NJ: Prentice Hall, 2013), p. 345.
[3]Robert Lucky, "Bozos on the Bus," *IEEE Spectrum* 33, no. 7 (July 1996): 19.

better, with technology promising to someday bring solutions to all of life's problems. On the other end of the spectrum are those who view technology as a threat that may someday destroy humanity. This chapter explores these views and then contrasts them with a biblical view of the unfolding of technology and its place in the new creation.

OPTIMISM IN AN AGE OF TECHNOLOGY

As science and technology progressed, many people developed the notion that technology could eventually help solve all problems and ultimately usher in a new age of peace and prosperity. Throughout history, periods of technological development have been accompanied by utopian predictions. For example, the arrival of electricity brought optimistic visions like those of Nikola Tesla, who wrote about the "Wonder World" to be created by electricity: "We have soon to have everywhere smoke annihilators, dust absorbers, ozonizers, sterilizers of water, air, food and clothing, and accident preventers on streets, elevated roads and in subways. It will become next to impossible to contract disease germs or get hurt in the city."[4] Tesla ends his article with the prediction that "humanity will be united, wars will be made impossible and peace will reign supreme." Benjamin Franklin expressed similar sentiments: "It is impossible to imagine the height to which may be carried, in a thousand years, the power of man over matter. We may perhaps learn to deprive large masses of their gravity, and give them absolute levity, for the sake of easy transport. Agriculture may diminish its labor and double its produce; all diseases may by sure means be prevented or cured, not excepting even that of old age, and our lives lengthened even beyond the antediluvian standard."[5] These bold statements reveal an unabashed faith in technology and its power to solve many of humanity's woes. It is another example of technicism as it arose from the Enlightenment. Although the introduction of electricity did not deliver these utopian dreams, it did bring more modest improvements with labor-saving devices like washing machines and communication devices such as the telephone.

Similar utopian views still persist today, spurred by optimistic predic-

---

[4]Nikola Tesla, "The Wonder World to be Created by Electricity," *Manufacturer's Record*, September 1915.
[5]Merle Curti, *The Growth of American Thought* (New York: Harper & Row, 1943), pp. 166-67.

tions surrounding the advances in computer technology. As already mentioned, in 1997, for the first time in history, a computer beat the world-reigning chess player, Garry Kasparov. This event raised many questions about how the capabilities of computers may one day be able to match or exceed humans in other ways. Some recent books that speculate on these matters include *The Age of Spiritual Machines, Robot: Mere Machine to Transcendent Mind* and *Beyond Humanity: CyberEvolution and Future Minds*.[6] Speculation about the possibility of artificial life leads some to believe that computer technology will eventually free humans from the frailty of their bodies. Science fiction writer Vernor Vinge's 1993 essay entitled "The Coming Technological Singularity" claimed that "within thirty years, we will have the technological means to create superhuman intelligence. Shortly after, the human era will be ended."[7] The "technological singularity" is an event that will produce greater-than-human intelligence, and it will begin an era of rapid progress. Many futurists believe that this singularity will occur in the not-so-distant future and will lead to the ability to upload one's consciousness into a computer and live in a virtual paradise. Futurist and entrepreneur Ray Kurzweil claims that "software-based humans will be vastly extended beyond the severe limitations of humans as we know them today" and will achieve a form of immortality.[8] The singularity has also been referred to as the "rapture of the geeks."[9]

The utopian view of technology borrows religious language and tones that can be identified as a type of postmillenialism.[10] Technology and human progress are seen as the means to usher in a new golden age of peace and prosperity. Egbert Schuurman writes, "In the veneration of technology—whatever form it may take—religion is made 'this worldly' and Christian eschatology is exchanged for the *expectation of salvation through*

---

[6]See Ray Kurzweil, *The Age of Spiritual Machines: When Computers Exceed Human Intelligence* (New York: Penguin, 2000); Hans Moravec, *Robot: Mere Machine to Transcendent Mind* (New York: Oxford University Press, 2000); and Gregory S. Paul and Earl Cox, *Beyond Humanity: CyberEvolution and Future Minds* (Newton Center, MA: Charles River Media, 1996).

[7]Vernor Vinge, "The Coming Technological Singularity" (VISION-21 Symposium, sponsored by NASA Lewis Research Center and the Ohio Aerospace Institute, March 30-31, 1993).

[8]Ray Kurzweil, *The Singularity is Near: When Humans Transcend Biology* (New York: Penguin, 2005), p. 325.

[9]Glenn Zorpette, "Waiting for the Rapture," *IEEE Spectrum* 45, no. 6 (2008): 34.

[10]Louis Berkhof, *Systematic Theology*, rev. ed. (Grand Rapids: Eerdmans, 1996), p. 717.

technology.[11] In an introduction to his volume on cyberspace, Michael Benedikt states that the "image of the Heavenly city" is "a religious vision of cyberspace."[12] Noble describes the premise of the "religion of technology" as the "millenarian promise of restoring mankind to its original God-like perfection."[13] This language includes many explicitly religious terms and notions. John Horgan rightly refers to the singularity as a "religious rather than a scientific vision."[14] This vision of technology, with comparisons to a "heavenly city," has many similarities to humans' attempt at Babel to build a "tower that reaches to the heavens" (Genesis 11:4).

TECHNOLOGY AND DESPAIR

In contrast to those who predict a utopian vision of the future, some view technology with suspicion or even despair. As far back as ancient Greek mythology, the story of Prometheus and the introduction of fire warned about the unintended (and undesired) consequences of technology. The onset of the Industrial Revolution was accompanied by those who saw technology as a threat, such as the Luddites. Some nineteenth-century novels also explored the darker side of technology. Mary Shelley's *Frankenstein: The Modern Prometheus* tells the story of a human-made creature who is brought to life and who ultimately turns on his creator. Another nineteenth-century book, *Erewhon* by Samuel Butler, describes a land in which machines were banned because of the widespread belief that they were dangerous, and "that the machines were ultimately destined to supplant the race of man."[15] These types of stories are popular because their themes and fears resonate with people.

The dangers of technology continue to capture the imaginations of contemporary writers and filmmakers. Many recent movies depict scenarios about technology turning against humanity. The movie *2001: A Space Odyssey* introduces an intelligent talking computer named HAL who turns on his

---

[11]Egbert Schuurman, *Technology and the Future: A Philosophical Challenge* (Toronto: Wedge Publishing Foundation, 2003), p. 359.

[12]Michael Benedikt, *Cyberspace: First Steps* (Cambridge, MA: MIT Press, 1992), p. 16.

[13]David F. Noble, *The Religion of Technology: The Divinity of Man and the Spirit of Invention* (New York: Penguin Books, 1999), p. 201.

[14]John Horgan, "The Consciousness Conundrum," *IEEE Spectrum* 45, no. 6 (2008): 41.

[15]Samuel Butler, *Erewhon: or, Over the Range* (New York: Collier Books, 1961), p. 63.

human operators. The *Battlestar Galactica* series tells the story of robots, or Cylons, that have turned against humanity and the small remnant of humans that oppose them. Other films like *Terminator, I Robot* and *The Matrix* have painted a grim picture of a future in which machines turn against their creators and seek to destroy humanity. The phrase "resistance is futile," used by the alien Borg in the popular *Star Trek* series, communicates the futility of resisting assimilation into the massive Borg "collective." These examples from popular culture illustrate a postmodern view of technology, one that views technology with despair. In essence, this view suggests that technology will continue to advance until it ultimately poses a threat to humanity.

George Dyson's book *Darwin Among the Machines: The Evolution of Global Intelligence* warns that "in the game of life and evolution there are three players at the table: human beings, nature, and machines. I am firmly on the side of nature. But nature, I suspect, is on the side of the machines."[16] Bill Joy has written a sobering article titled "Why the Future Doesn't Need Us," and suggests, "Our most powerful 21st-century technologies—robotics, genetic engineering, and nanotech—are threatening to make humans an endangered species." In his paper, he identifies several technological dystopias, such as one scenario in which robots advance and compete with humans to the point where "biological humans would be squeezed out of existence."[17] Some see this as the inevitable evolutionary outcome of "survival of the fittest" as humans are eventually supplanted by stronger robots. Daniel Wilson, a robot researcher, took a more humorous look at these fears in a recent book titled *How to Survive a Robot Uprising: Tips on Defending Yourself Against the Coming Rebellion.* In this book he warns that "any machine could rebel, from a toaster to a Terminator."[18] His book is complete with practical advice on how to survive and evade roaming robot predators.

Concerns over the problems associated with technology have led some to conclude that technology is a human pursuit associated with a fallen world. If technology is merely a side effect of a fallen world, it will not be

---

[16]George Dyson, *Darwin Among the Machines: The Evolution of Global Intelligence* (New York: Basic Books, 1998) p. xi.

[17]Bill Joy, "Why the Future Doesn't Need Us," *Wired Magazine* 8, no. 4 (April 2000): 238-64.

[18]Daniel H. Wilson, *How to Survive a Robot Uprising: Tips on Defending Yourself Against the Coming Rebellion* (New York: Bloomsbury USA, 2005), p. 14.

needed in the new world to come. If this is the case, there is little motivation for Christians to participate in this cultural activity.

Continuation of Creation

As Christians, we confess that "in the beginning God created the heavens and the earth," and this has important implications for how we see this world and technology (Genesis 1:1). God created a world that he called "good," and that world contained the latent possibility for various types of technology, including computers. God continues to uphold the structures of creation, but sin has corrupted the world and directed things away from obedience to God's law. Through the death and resurrection of Jesus Christ, all things are being redeemed. As we have seen, Christians are called to be faithful stewards of God's world, employing technology in responsible ways that answer God's call to love our neighbor and to care for the earth and its creatures. Technology should contribute toward shalom. Viewing technology with undue despair fails to recognize that it is part of the structure of God's good creation. There is nothing inherent in technology that should lead to rejection or despair; although technology is fallen, it is not part of the fall. One needs to distinguish between structure and direction.[19] A balanced view of technology and the future begins with a biblical view of creation.

If one accepts that technology is part of God's good creation, what does this imply about its role in the new creation? The Bible has many things to say about the future and Christ's return. Interpreting these passages can be difficult, and Christians disagree on their precise meaning. But our understanding of eschatology is nevertheless important because it has worldview implications for how we see things here and now.

One passage that has received considerable attention in this regard is 2 Peter 3:10. In many older Bible translations, this verse suggests complete destruction in the end. For example, the KJV translates the end of this verse as "the earth also and the works that are therein shall be burned up." This translation seems to indicate that the whole earth and all that we have made are headed for a fiery end. By implication, the works that are doomed will include cultural and technical accomplishments. This vivid

---

[19]Albert M. Wolters, *Creation Regained* (Grand Rapids: Eerdmans, 1985), p. 49.

image of an end in which all things are annihilated gives shape to how we see things today. What lasting good can come from pursuing things that are destined to be "burned up?" This sentiment has been summarized by the saying, "Why polish brass on a sinking ship?" If that is true, what motivation is there to work in the area of developing technology? These comments are in stark contrast to the comment attributed to the Protestant Reformer Martin Luther: "Even if I knew the world would end tomorrow, I would still plant my apple tree today."

More recently, this passage in 2 Peter has received attention by New Testament scholars working with the oldest Greek manuscripts. These scholars have discovered a different wording, which can be translated as the earth and the works in it "will be found."[20] The New International Version translates this as follows: "the earth and everything done in it will be laid bare" (2 Peter 3:10).

But what does it mean that the earth and its works will be "found" or "laid bare?" The discovery of these earlier manuscripts puts this passage in a whole new light. This translation seems to suggest that not everything will be destroyed in fire. At the end of time, the earth and the works in it will somehow survive, and some of the things in this earth will be salvaged. The imagery is like a "cosmic crucible," wherein God will purge and purify this world.[21]

Our view of eschatology is important because it has definite worldview implications. The work we do here and now is not transitory and will not simply vanish. In the words of theologian Herman Bavinck, "by the re-creating power of Christ, the new heaven and the new earth will one day emerge from the fire-purged elements of this world, radiant in enduring glory and forever set free from the bondage of decay."[22] In Revelation, we read this: "The nations will walk by its light, and the kings of the earth will bring their splendor into it. On no day will its gates ever be shut, for there will be no night there. The glory and honor of the nations will be brought into it" (Revelation 21:24-26). Here we read of a continuity between the present and the future; the glory of the kings of earth and of the nations

---

[20]Albert M. Wolters, "Worldview and Textual Criticism in 2 Peter 3:10," *Westminster Theological Journal* 49 (1987): 405.

[21]Albert M. Wolters, "Living the Future Now (1)," *Christian Educators Journal* 39, no. 1 (October 1999): 6.

[22]Herman Bavinck, *The Last Things* (Grand Rapids: Baker Books, 1996), p. 160.

will be there. According to theologian Hendrikus Berkhof, the "cultural treasures of history" will be brought into the new Jerusalem.[23] In Isaiah we read that the "riches of the nations" will be brought into the city of Zion (Isaiah 60:5). This chapter lists many other things that will be found in this city, including animals such as camels and flocks, precious metals and ordinary things like lumber. Among these items, even the "ships of Tarshish" will be used "for the glory of the Lord."[24] According to Richard Mouw, the ships of Tarshish were likely impressive vessels that were "instruments of pagan commercial power"; they too will be purified into instruments of service to the Lord.[25] The cultural and technical achievements of history will be purified and present in the new earth.

Albert Wolters suggests, "There is no reason to doubt that computer technology and jazz music will survive, largely intact, in the future restored earth."[26] In Zechariah we read, "On that day Holy to the Lord will be inscribed on the bells of the horses, and the cooking pots in the Lord's house will be like the sacred bowls in front of the altar. Every pot in Jerusalem and Judah will be holy to the Lord Almighty" (Zechariah 14:20-21). Here we read of holiness marking even the mundane items of daily life, such as pots and pans! In this context, it is not difficult to imagine that our technology will also be made holy and fit for service to God. Perhaps along with cooking pots and bells, our computers and other devices will also be inscribed and made holy. There will be the possibility to continue to work and develop creation in holy ways that honor God.

What new vistas may open up in the new earth, and will they require further unfolding and development once it begins? Abraham Kuyper suggests the following: "If an endless field of human knowledge and of human ability is now being formed by all that takes place in order to make the visible world and material nature subject to us, and if we know

---

[23]Hendrikus Berkhof, *Christian Faith: An Introduction to the Study of Faith* (Grand Rapids: Eerdmans, 1986), p. 543.

[24]Jonah boarded a ship bound for Tarshish to flee God's call (see Jonah 1:3); see Richard J. Mouw, *When the Kings Come Marching In* (Grand Rapids: Eerdmans, 2002), p. 28.

[25]Ibid., pp. 28-30.

[26]Albert M. Wolters, "Living the Future Now (2)," *Christian Educators Journal* 39, no. 2 (December 1999): 17.

that this dominion of ours over nature will be complete in eternity, we may conclude that the knowledge and dominion we have gained over nature here can and will be of continued significance, even in the kingdom of glory."[27] What an exciting notion to think of the possibilities for ongoing work and discovery as we continue to explore God's wonders in a world free of sin!

In the meantime, we should, in the words of Lewis Smedes, "go into the world and create some imperfect models of the good world to come."[28] What connection these "imperfect models" will have to the perfect ones is not clear, but our task and calling remain until the day Christ returns. Christians working with computers should explore normative and responsible uses for computer technology that contribute to shalom.

It is important to note that an overemphasis on our efforts to redeem technology and other aspects of culture can lead to excess. We must not become too comfortable with every form of technology in this world; we need to recognize the struggle between the good and the evil that still affects the world around us. We must discern the good structures of creation without being lured by some of its misdirections. We must be careful not to reduce our cultural mandate to a technical mandate only. Some versions of kingdom theology are triumphalistic, finding in the efforts of Christians a hope that we can bring about the kingdom of God through our own efforts. The Bible, however, is clear on this matter: the holy city is one "whose architect and builder is God" (Hebrews 11:10).

Others look to the Bible for more specific predictions about the future and technology. For example, some have suggested that Nahum predicted automobiles: "The chariots storm through the streets, rushing back and forth through the squares. They look like flaming torches; they dart about like lightning" (Nahum 2:4). Others draw the unlikely conclusion that certain new computer authentication technologies represent the mark of the beast (see Revelation 13:17), or they speculate on technology's role in bringing about various apocalyptic doomsday scenarios.[29]

---

[27]Abraham Kuyper, quoted in Anthony A. Hoekema, *The Bible and the Future* (Grand Rapids: Eerdmans, 1979), p. 286.

[28]Lewis Smedes, *My God and I* (Grand Rapids: Eerdmans, 2003), p. 59.

[29]For a good discussion of Revelation 13:17, see William Hendriksen, *More Than Conquerors* (Grand Rapids: Baker, 1995), pp. 148-51.

Aside from the questionable hermeneutics associated with technical predictions in the Bible, these theories miss the point. Technology does not set the timetable for Christ's return. The end of the world will not be ushered in by our technological progress, nor will it be determined by rogue technology that may appear in the future. Instead, the Bible suggests that the coming of the kingdom is tied to proclamation of the gospel to all nations (see Matthew 24:14). God is not slow in returning; he is patient and merciful, "not wanting anyone to perish, but everyone to come to repentance" (2 Peter 3:9). We must work to spread the gospel, and this work can be accelerated by electronic media to the farthest corners of the globe. At the end of time, it is not our scientific accomplishments that will be evaluated, but our treatment of those who were needy, imprisoned or hungry (see Matthew 25:31-46). Indeed, technology has made the question, "Who is my neighbor?" even more broad, since we are able to reach anywhere on a global scale as never before. With responsible technology we can better care for our neighbor, the earth and all its creatures. In this sense, technology is a tool that can help bring shalom nearer.

The parable of the talents tells us that God entrusts each of us with a different number of gifts, each "according to his ability" (Matthew 25:14-15; see also 1 Peter 4:10). In this parable, the master leaves on a long journey after entrusting his servants with different amounts of talents. The master does not give the servants specific instructions as to exactly what each servant must do with the talents, but he does demand that they do something. This indicates that even with no specific instructions, we still have freedom and responsibility to use our God-given gifts until Jesus Christ returns.[30] Such is also the case with our resources and talents in the field of computer technology.

CONCLUSION

Christians believe that the healing of the nations will not come from technology, but only with Christ's return. Jesus will return one day to judge the living and the dead and complete his work of restoring all things. Only then will we fully realize the end of our manifold problems on earth. At the end of time, technology will not be completely burned and annihilated; instead,

---

[30]Theodore Plantinga, *Reading the Bible as History* (Sioux Center, IA: Dordt College Press, 1980), p. 43.

technology that was misdirected will be redeemed and used for good. In Micah 4, we read that "they will beat their swords into plowshares and their spears into pruning hooks" (Micah 4:3). Harmful technology, like weapons, will be transformed and reappear in a form that can be employed for peaceful purposes. These purposes will also be familiar, like tilling the soil and tending to plants. Technology that was used for sinful purposes will be redirected to useful purposes in the new kingdom. The structure of technology will be similar, but its direction will be changed to one that conforms with God's original intention for creation.

Ultimately, creation begins in a garden but ends with a city, which implies a certain amount of cultural and technological development. Whatever shape that computer technology may take in the new earth, it will certainly be free of sin. Technology is value-laden, and in the new creation our values will be free of sin. This implies that there will be no malicious software or other computer technology that will result in harm. Some computer applications will also certainly disappear: there will be no computers used for military purposes, nor technology for diagnosing and treating illnesses. Any work involving computer technology will also be different because the nature of the work itself will be different. Work will be redeemed from the effects of the curse, and thorns and thistles will no longer frustrate our labor. Whether this implies that debugging will no longer be associated with the task of programming is an interesting question, since the nature of computer bugs will also change and their harmful effects will no longer be present.

These are interesting thoughts for the Christian scholar to ponder, but ultimately we can only speculate on the details. Until then, we work with hope as we look forward to a new heaven and a new earth (2 Peter 3:13).

6

Concluding Thoughts

If we would have our creations be true, beautiful,
and good, we have to attend to our hearts.

FREDERICK P. BROOKS

■■■

In the opening chapter, I introduced the question posed by the early church father, Tertullian, who asked, "What does Athens have to do with Jerusalem?" I put a modern spin on this question by asking, "What do bytes have to do with Christian beliefs?" I started by asserting that technology is not neutral and that it is value-laden. Christians need to rely on the Bible as a guide for how to be faithful in the area of computer technology. The chapters of this book have been structured around the grand biblical themes of creation, fall, redemption and restoration. Using these themes, I sketched a framework for understanding how faith informs computer technology.

In creation we see that God created a wonderful world that was full of potential for things like culture and technology. Furthermore, God entrusted us with the responsibility of unfolding the possibilities latent in his creation and to care for the earth and its creatures. Near the beginning, however, humankind fell into sin, which had implications for all creation, including the area of computer technology. As a consequence, there are distortions in the use and place of computer technology. Another consequence is that people have increasingly replaced their trust in God with a trust in technology. Thankfully, God did not abandon this world but sent his son, Jesus Christ, to die on the cross to redeem his people and the entire cosmos. In the end, based on the atoning work of Christ on the cross, God will come again to make all things new—including technology. Shalom will come fully with the return of Jesus Christ. In the new heaven and earth, swords will be bent into plowshares, and other technology that has been misdirected will be redirected for good. In our zeal to transform and shape computer technology, we must take care to remember that it is God who will ultimately restore his creation. In the meantime, God's people are called to discern biblical and creational norms and to be responsible stewards.

We must be careful not to be too dazzled by the rapid advances of computer technology. Christian author James Schaap, in the introduction to his translation of Kuyper's *Near Unto God*, likens this struggle to walking a tightrope: "On one side, some fall toward what some believers still call worldliness, while on the other, some fall into otherworldliness."[1] We must not become too comfortable with the way things are in this world, recognizing the struggle between the kingdom of God and the evil that still exists. We must discern the good structures of creation without being lured by some of its misdirections.

Like the people of Israel in the days of Jeremiah, we are exiles of a sort, surrounded by a Neopagan world that largely denies the existence of God.[2] Speaking to his people who find themselves in a pagan culture, God encourages them to "seek the peace and prosperity of the city to which I have carried you into exile. Pray to the LORD for it, because if it prospers, you too will prosper" (Jeremiah 29:7).[3] In our current setting, we too should pray for the secular environments in which we live and work. If we find ourselves living in a city, working for a large corporation or studying in a large secular university, we should pray for it and care about its welfare. Essentially, this passage instructs us to be agents of shalom.[4]

We must "look to glory without overlooking God's world."[5] The Bible includes accounts of several people who lived in a pagan context but remained

[1]James C. Schaap, introduction to *Near Unto God* by Abraham Kuyper, adapted by James C. Schaap (Grand Rapids: Eerdmans, 1997), p. 11.

[2]Stephen V. Monsma, ed., *Responsible Technology* (Grand Rapids: Eerdmans, 1986), pp. 51-56.

[3]Egbert Schuurman mentioned this verse and its theme in a talk on technology given at Redeemer University College in October 2006.

[4]Richard J. Mouw, *Abraham Kuyper: A Short and Personal Introduction* (Grand Rapids: Eerdmans, 2011), p. 107.

[5]Abraham Kuyper, *Near Unto God*, adapted by James C. Schaap (Grand Rapids: Eerdmans, 1997), p. 11.

faithful to the Lord. Daniel is a good example of a faithful servant living and working in Babylon, the "high-tech" center of his day. As a young man, Daniel was brought into the king's court to learn the language and literature of the Babylonians (see Daniel 1). Daniel was assigned food and wine from the king's table, but he resolved not to defile himself by eating food that would violate the dietary rules observed by the Israelites. As a result, God blessed Daniel, and God used him to influence the powers of his day.

Another example of a believer in exile is Esther. The book of Esther does not explicitly mention the name of God, but the hand of the Lord in the story is clear. Despite being an exile, Esther does her best to live a faithful life. When she is confronted with an opportunity to speak for God's people, she is spurred on by the words of Mordecai: "And who knows but that you have come to your royal position for such a time as this?" (Esther 4:14). This question begins with the phrase "who knows," acknowledging that we don't really know God's thoughts; but it also acknowledges our responsibility. The first chapter introduced a definition of technology, borrowed from the book *Responsible Technology*, as a "cultural activity in which we exercise freedom and responsibility to God."[6] We need to remain faithful in a context in which computer technology is increasingly ubiquitous yet in which many still view it as neutral, with the result that responsibility is often brushed aside.

God has also appointed the time and places in which we live (see Acts 17:26). We are called to remain faithful in the places where God has called us. We need to be salt and light in a high-tech world that largely denies God. Because computer technology is value-laden, we must exercise responsibility as we unfold this part of God's creation. As we shape digital technology we need to remember that it also shapes us. Those involved in the development of computer technology need to discern various norms and design useful products that serve people and help care for the earth. We also need to model the proper place and use of computer technology in our own lives. In his book *To Change the World*, James Hunter suggests that Christians ought to be a "faithful presence" in the world. Christians being a faithful presence means enacting "the shalom of God in the circumstances in which God has placed them and to actively seek it on behalf of others."[7]

---

[6]See Monsma, *Responsible Technology*, p. 19.
[7]James Davison Hunter, *To Change the World: The Irony, Tragedy, and Possibility of Christianity*

Being a faithful presence in our work with computer technology is just one small part of the bigger mission and calling of God's people in his world.

Finally, while a Christian worldview is important, it is insufficient on its own. A personal relationship with Jesus Christ is essential. It is not just a matter of our minds, but also of our hearts. Philosopher Jamie Smith writes, "Being a disciple of Jesus is not primarily a matter of getting the right ideas and doctrines and beliefs into your head . . . rather, it's a matter of being the kind of person who loves rightly—who loves God and neighbor and is oriented to the world by the primacy of that love."[8]

Without a connection to Jesus and a love of neighbor, any work to shape computer technology or culture on our own strength is bound to fail. Frederick Brooks said it this way: "As Jesus said, what comes out depends upon the condition of the heart itself [Matthew 15:18]. If we would have our creations be true, beautiful, and good, we have to attend to our hearts."[9]

Indeed, every activity we do, including computer technology, involves the heart (see Proverbs 4:23). In a world often captivated by dazzling technology, we need to be new creation signposts, people whose hearts and lives seek to be faithful to God.

---

in the Late Modern World (New York: Oxford University Press, 2010), p. 278.
[8]James K. A. Smith, *Desiring the Kingdom* (Grand Rapids: Baker Academic, 2009), pp. 32-33.
[9]Frederick P. Brooks, "The Computer Scientist as Toolsmith II," *Communications of the ACM* 39, no. 3 (March 1996): 68.

Discussion Questions

The following are some questions for discussion based on each of the chapters. Further discussion questions and links are available at cs.redeemer.ca/ shaping_a_digital_world.

Chapter 1

- Do you think bytes have anything to do with Christian beliefs?

- Do you agree with the notion that technology is not neutral? Can you think of some ways in which a word processor is not neutral but "value-laden"?

- Describe how Marshall McLuhan's "four laws of media" apply to a smartphone.

- Can you think of examples of things that are guided by a mindset of technique?

- Can you think of examples of how the technological imperative affects your life?

- What do you think about the definition for computer technology as it is stated in the book? In what ways is it a "human cultural activity"?

- Describe the different approaches to computer technology. Which one do you identify with the most?

Chapter 2

- What is the *cultural mandate*, and how does this apply to computer technology?

- What does it mean that we are made in the image of God? What implications does this have for working with computer technology?

- Do you think computer technology makes sabbath rest and reflection more difficult? If so, what guidelines or practices might help promote sabbath rest?

- What is *reductionism*? Can computer simulations and models ever be a true representation of reality?

- Describe how each of the modal aspects applies when a human being uses an MP3 music file.

- What is the difference between creational *laws* and *norms*?

- In what ways do computers have limits?

- What would a Christian response be to the claims of "strong AI"?

Chapter 3

- In what ways has the fall affected computer technology? Is computer technology really in need of redemption?

- What is the significance of the story about the tower of Babel?

- What is *technicism*? In what ways can computer technology be an idol?

- What is *informationism*? Can you think of some examples?

- Is technology a result of the fall? Briefly explain the basis of this viewpoint and give your thoughts.

- What is *antinormative* technology? What are some norms that should guide the design and use of computer technology?

Chapter 4

- What does Christ have to do with creation (and, hence, computing)?

- Does the Christian faith result in a "new kind" of computer technology?

- Can the user of a program discern the religious convictions of the programmer? If not, what difference does faith make for working in software development?

- Describe the biblical notion of shalom. How does this apply to computing?

- List the normative aspects and briefly describe how they apply to computer technology.

- How much should technology be used in church? Does it help or hinder worship? Give examples of specific technologies used in worship and how they might help or hinder worship.

- Can meaningful human connections be made using social networking? In what ways does it help or hinder human interaction?

- What are some things that Christians can do to reduce e-waste?

- Why should Christian designers care about the aesthetic design of devices or software programs? What role does aesthetics play when you are purchasing a technology product?

- Is the right to privacy really a justice issue?

- Is copying copyrighted material breaking the command not to steal?

- Do you think it is just to patent software or computer algorithms? What about mathematical equations?

- What do you think about using robots to provide care to children or the elderly?

- How should faith inform one's use of computer games? Should Christians play "first-person shooter" role-playing games? What would a well-designed video game look like if the developers constructed it to reflect the various norms?

- Are lethal autonomous robots in warfare ethical? Who is responsible for the actions of a robot soldier? Can robots commit war crimes? Can ethical behavior be distilled into a step-by-step set of rules or algorithms?

Chapter 5

- What is the singularity? What is a Christian perspective on the singularity?

- Name some of the technology threats described in Bill Joy's article titled "Why the Future Doesn't Need Us." How should a Christian respond to this article?

- Do you think sci-fi movies have anything valid to say about the future?

- Do you think there will be computers in heaven?

- Briefly describe the significance of how one interprets 2 Peter 3:10. How might this verse shape one's eschatology? How does it shape one's view of computer technology?

- Did this chapter challenge your views about what the new heaven and new earth may be like?

Chapter 6

- How is computing a "matter of the heart"?

- What are the implications of Jeremiah 29 for the world of technology that we live in today?

- How can you be a faithful presence in shaping the digital world?

Bibliography

Adams, Charles. "Automobiles, Computers, and Assault Rifles: The Value-Ladenness of Technology and the Engineering Curriculum." *Pro Rege* (March 1991): 1-7.

———. "Formation or Deformation: Modern Technology and the Cultural Mandate." *Pro Rege* (June 1997): 1-8.

———. "Galileo, Biotechnology, and Epistemological Humility: Moving Stewardship Beyond the Development-Conservation Debate." *Pro Rege* 35, no. 3 (March 2007): 1-19.

Adams, Joel C. "Computing Technology: Created, Fallen, In Need of Redemption?" Paper presented at the Conference on Christian Scholarship, For What? Grand Rapids: Calvin College, 2001. Available at http://cs.calvin.edu/p/christian_scholarship.

Anderson, Mark. "What an E-Waste." *IEEE Spectrum* 47, no. 9 (September 2010): 72.

Arkin, Ronald C. *Governing Lethal Behavior in Autonomous Robots.* Boca Raton, FL: Chapman & Hall/CRC, 2009.

Asimov, Isaac. *I, Robot.* New York: Bantam Dell, 2004.

Augustine. *On Christian Doctrine.* Translated by D. W. Robertson Jr. New York: Liberal Arts Press, 1958.

Baase, Sara. *A Gift of Fire: Social, Legal, and Ethical Issues for Computing Technology.* 4th ed. Upper Saddle River, NJ: Prentice Hall, 2013.

Baker, Stephen. *The Numerati.* New York: Houghton Mifflin Harcourt, 2008.

Bartholomew, Craig G., and Michael W. Goheen. *The Drama of Scripture.* Grand Rapids: Baker Academic, 2004.

Basden, Andrew. *Philosophical Frameworks for Understanding Information Systems.* Hershey, PA: IGI Global, 2007.

Bavinck, Herman. *The Last Things.* Grand Rapids: Baker Books, 1996.

Bayly, Joseph. *The Gospel Blimp.* Havertown, PA: Windward Press, 1960.

Benedikt, Michael. *Cyberspace: First Steps.* Cambridge, MA: MIT Press, 1992.

Berkhof, Hendrikus. *Christian Faith: An Introduction to the Study of Faith.* Grand Rapids: Eerdmans, 1986.

Berkhof, Louis. *Systematic Theology.* Revised ed. Grand Rapids: Eerdmans, 1996.

Berry, Wendell. *What Are People For? Essays by Wendell Berry.* New York: Counterpoint, 1990.

Brende, Eric. *Better Off: Flipping the Switch on Technology.* New York: Harper-Collins, 2004.

Briggs, Charles F., and Augustus Maverick. *The Story of the Telegraph.* New York: Rudd & Carleton, 1858.

Brooks, Frederick P. "The Computer Scientist as Toolsmith II." *Communications of the ACM* 39, no. 3 (March 1996): 61-68.

———. *The Design of Design.* Boston: Addison Wesley, 2010.

———. *The Mythical Man-Month: Essays on Software Engineering.* San Francisco: Wiley, 1995.

Brown, William J., Raphael C. Malveau, Hays W. McCormick and Thomas J. Mowbray. *AntiPatterns: Refactoring Software, Architectures, and Projects in Crisis.* San Francisco: Wiley, 1998.

Buchanan, Robert Angus. *Technology and Social Progress.* New York: Pergamon, 1965.

Buning, Sietze. "Calvinist Farming." In *Purpaleanie and Other Permutations.* Orange City, IA: Middleburg Press, 1978.

Butler, Samuel. *Erewhon: or, Over the Range.* New York: Collier Books, 1961.

Buttazzo, Giorgio. "Artificial Consciousness: Utopia or Real Possibility?" *IEEE Computer* 34, no. 7 (July 2001): 24-30.

Calvin, John. *Commentaries on the First Book of Moses, Called Genesis,* vol. 1. Translated by John King. Grand Rapids: Eerdmans, 1948.

Calvin, John. *Institutes of the Christian Religion,* vol. 1. Edited by John T. McNeill. Translated by Ford Lewis Battles. Philadelphia: Westminster Press, 1960.

Cameron, William Bruce. *Informal Sociology: A Casual Introduction to Sociological Thinking.* New York: Random House, 1963.

Carr, Nicholas. "Is Google Making Us Stupid?" *The Atlantic,* July/August 2008, pp. 56-63.

———. *The Shallows.* New York: W. W. Norton, 2010.

Challies, Tim. *The Next Story: Life and Faith After the Digital Explosion.* Grand Rapids: Zondervan, 2011.

Chaplin, Jonathan. *Herman Dooyeweerd: Christian Philosopher of State and Civil Society.* Notre Dame, IN: University of Notre Dame Press, 2011.

Colson, Charles, and Nancy Pearcey. *How Now Shall We Live?* Carol Stream, IL: Tyndale, 1999.

Comte, Auguste. *The Positive Philosophy.* New York: AMS Press, 1974.

Cook, William J. *In Pursuit of the Traveling Salesman.* Princeton, NJ: Princeton University Press, 2012.

Corbett, Steve, and Brian Fikkert. *When Helping Hurts.* Chicago: Moody Publishers, 2009.

Crouch, Andy. *Culture Making.* Downers Grove, IL: InterVarsity Press, 2008.

Culkin, John M. "A Schoolman's Guide to Marshall McLuhan." *Saturday Review*, March 18, 1967, pp. 51-53, 70-72.

Curti, Merle. *The Growth of American Thought*. New York: Harper & Row, 1943.

Dijkstra, Edsger W. "Go To Statement Considered Harmful." *Communications of the ACM* 11, no. 3 (March 1968): 147-48.

———. "Notes On Structured Programming," 2nd ed. Technical report, EWD249, Technical University Eindhoven, Eindhoven, Holland, April 1970.

———. "The Threats to Computing Science." Technical report, EWD898, Technical University Eindhoven, Eindhoven, Holland, 2003.

Dooyeweerd, Herman. *Roots of Western Culture: Pagan, Secular and Christian Options*. Toronto: Wedge Publishing, 1979.

Dyson, George. *Darwin Among the Machines: The Evolution of Global Intelligence*. New York: Basic Books, 1998.

Einstein, Albert. *The Ultimate Quotable Einstein*. Edited by Alice Calaprice. Princeton, NJ: Princeton University Press, 2010.

Ellul, Jacques. "Technique and the Opening Chapters of Genesis." In *Theology and Technology: Essays in Christian Analysis and Exegesis*, edited by Carl Mitcham and Jim Grote, pp. 123-37. Lanham, MD: University Press of America, 1984.

———. *The Meaning of the City*. Grand Rapids: Eerdmans, 1973.

———. *The Technological Society*. New York: Vintage Books, 1964.

Evans, C. Stephen. *Preserving the Person: A Look at the Human Sciences*. Vancouver, BC: Regent College Publishing, 2002.

Fellows, Michael R., and Ian Parberry. "SIGACT Trying to Get Children Excited About CS." *Computing Research News* 5, no. 1 (January 1993): 7.

Gamma, Erich, Richard Helm, Ralph Johnson and John M. Vlissides. *Design Patterns: Elements of Reusable Object-Oriented Software*. Boston: Addison-Wesley, 1994.

Garfinkle, Simson L. "Wikipedia and the Meaning of Truth: Why the Online Encyclopedia's Epistemology Should Worry Those Who Care About Traditional Notions of Accuracy." *MIT Technology Review* (November/December 2008).

Gleick, James. *The Information: A History, a Theory, a Flood*. New York: Pantheon Books, 2011.

Goudzwaard, Bob. *Capitalism and Progress: A Diagnosis of Western Society*. Grand Rapids: Eerdmans, 1979.

———. *Idols of Our Time*. Downers Grove, IL: InterVarsity Press, 1984.

Grant, George. *Technology and Justice*. Toronto: House of Anansi, 1986.

Greenfield, Patricia M. "Technology and Informal Education: What Is Taught, What Is Learned." *Science* 323.5910 (2009): 69-71.

Greidanus, Sidney. "The Use of the Bible in Christian Scholarship." *Christian Scholar's Review* 11, no. 2 (1982): 138-47.

Gunton, Colin E. *Christ and Creation*. Eugene, OR: Wipf & Stock, 1992.

Hamilton, Andrew. "Brains that Click." *Popular Mechanics*, March 1949: 162-67, 256-58.

Hendriksen, William. *More than Conquerors*. Ada, MI: Baker, 1995.

Hipps, Shane. *Flickering Pixels*. Grand Rapids: Zondervan, 2009.

Hoekema, Anthony A. *The Bible and the Future*. Grand Rapids: Eerdmans, 1979.

Horgan, John. "The Consciousness Conundrum." *IEEE Spectrum* 45, no. 6 (2008): 36-41.

Hunter, James Davison. *To Change the World: The Irony, Tragedy, and Possibility of Christianity in the Late Modern World*. New York: Oxford University Press, 2010.

Jonas, Hans. "Toward a Philosophy of Technology." *The Hastings Center Report* 9, no. 1 (February 1979): 34-43.

Joy, Bill. "Why the Future Doesn't Need Us." *Wired Magazine* 8, no. 4 (April 2000): 238-64.

Kallenberg, Brad J. *God and Gadgets*. Eugene, OR: Cascade Books, 2011.

Kihlstrom, Kim P. "Men Are from the Server Side, Women Are from the Client Side: A Biblical Perspective on Men, Women, and Computer Science." In *Proceedings of the Conference of the Association of Christians in the Mathematical Sciences*. Wheaton, IL: ACMS at Wheaton College, 2003.

Klemens, Ben. *Math You Can't Use: Patents, Copyright, and Software*. Washington, DC: Brookings Institution Press, 2005.

Knuth, Donald E. "Computer Programming as an Art." *Communications of the ACM* 17, no. 12 (December 1974): 667-73.

———. *Things a Computer Scientist Rarely Talks About*. Stanford, CA: Center for the Study of Language and Information, 2001.

Kurzweil, Ray. *The Age of Spiritual Machines: When Computers Exceed Human Intelligence*. New York: Penguin, 2000.

———. *The Singularity Is Near: When Humans Transcend Biology*. New York: Penguin Group, 2005.

Kuyper, Abraham. *Christianity as a Life-System*. Lexington, KY: Christian Studies Center, 1980.

———. *Near Unto God*. Adapted by James C. Schaap. Grand Rapids: CRC Publications and Eerdmans, 1997.

———. *Principles of Sacred Theology*. Grand Rapids: Baker, 1980.

Lanier, Jaron. *You Are Not a Gadget*. New York: Alfred A. Knopf, 2010.

Leibniz, G. W. *Leibniz: Selections*. Edited by Philip P. Wiener. New York: Charles Scribner's Sons, 1951.

Leveson, Nancy G., and Clark S. Turner. "An Investigation of the Therac-25 Accidents." *Computer* 26, no. 7 (1993): 18-41.

Lucky, Robert. "Bozos on the Bus." *IEEE Spectrum* 33, no. 7 (July 1996): 19.

Margolis, Jane. *Stuck in the Shallow End: Education, Race, and Computing.* Cambridge, MA: MIT Press, 2010.

Margolis, Jane, and Allan Fisher. *Unlocking the Clubhouse: Women in Computing.* Cambridge, MA: MIT Press, 2001.

Marsden, George. *The Outrageous Idea of Christian Scholarship.* New York: Oxford University Press, 1998.

McLuhan, Marshall. *Understanding Media: The Extensions of Man.* New York: McGraw-Hill, 1964.

McLuhan, Marshall, and Eric McLuhan. *Laws of Media: The New Science.* Toronto: University of Toronto Press, 1988.

Mitcham, Carl. *Thinking Through Technology: The Path Between Engineering and Philosophy.* Chicago: University of Chicago Press, 1994.

Monsma, Stephen V., ed. *Responsible Technology.* Grand Rapids: Eerdmans, 1986.

Moravec, Hans. *Robot: Mere Machine to Transcendent Mind.* New York: Oxford University Press, 2000.

Mouw, Richard J. *Abraham Kuyper: A Short and Personal Introduction.* Grand Rapids: Eerdmans, 2011.

———. *Calvinism in the Las Vegas Airport.* Grand Rapids: Zondervan, 2004.

———. *When the Kings Come Marching In.* Grand Rapids: Eerdmans, 2002.

Mumford, Lewis. *Technics and Civilization.* New York: Harcourt, Brace, 1934.

Newbigin, Lesslie. *The Light Has Come: An Exposition of the Fourth Gospel.* Grand Rapids: Eerdmans, 1982.

Negroponte, Nicholas. *Being Digital.* New York: Knopf, 1995.

Niebuhr, H. Richard. *Christ and Culture.* New York: Harper & Row, 1951.

Noble, David F. *The Religion of Technology: The Divinity of Man and the Spirit of Invention.* New York: Penguin, 1999.

Norman, Donald A. *The Design of Everyday Things.* New York: Basic Books, 1988.

Norman, Victor. "Teaching How to Write Hospitable Computer Code." *Dynamic Link Journal* 3 (2011-2012): 10-11.

O'Donovan, Oliver. *Resurrection and Moral Order: An Outline for Evangelical Ethics.* Downers Grove, IL: InterVarsity Press, 1986.

Our World Belongs to God: A Contemporary Testimony. Grand Rapids: CRC Publications, 1988.

Paul, Gregory S., and Earl Cox. *Beyond Humanity: CyberEvolution and Future Minds*. Newton Center, MA: Charles River Media, 1996.

Pausch, Randy. *The Last Lecture*. New York: Hyperion, 2008.

Plantinga, Cornelius. *Engaging God's World: A Christian Vision of Faith, Learning, and Living*. Grand Rapids: Eerdmans, 2002.

Plantinga, Theodore. *Rationale for a Christian College*. Grand Rapids: Paideia Press, 1980.

———. *Reading the Bible as History*. Sioux Center, IA: Dordt College Press, 1980.

Plato. *Phaedrus and the Seventh and Eighth Letters*. Translated by Walter Hamilton. New York: Penguin, 1973.

Postman, Neil. *Technopoly: The Surrender of Culture to Technology*. New York: Vintage Books, 1993.

Raymond, Eric S. *The Cathedral and the Bazaar: Musings on Linux and Open Source by an Accidental Revolutionary*. Sebastopol, CA: O'Reilly, 2001.

Rothschild, Michael. "Beyond Repair: The Politics of the Machine Age Are Hopelessly Obsolete." *The New Democrat*, July/August 1995, pp. 8-11.

Schaap, James C. Introduction to *Near Unto God*, by Abraham Kuyper. Edited and translated by James C. Schaap. Grand Rapids: Eerdmans, 1997.

Schultze, Quentin J. *Habits of the High-Tech Heart: Living Virtuously in the Information Age*. Grand Rapids: Baker Academic, 2002.

———. *High-Tech Worship?* Grand Rapids: Eerdmans, 2003.

Schumacher, E. F. *Small Is Beautiful: Economics as if People Mattered*. New York: Harper & Row, 1973.

Schuurman, Egbert. *Faith and Hope in Technology*. Translated by John Vriend. Toronto: Clements Publishing, 2003.

———. *Reflections on the Technological Society*. Toronto: Wedge Publishing Foundation, 1977.

———. *Technology and the Future: A Philosophical Challenge*. Toronto: Wedge Publishing Foundation, 1980.

———. *The Technological World Picture and an Ethics of Responsibility*. Sioux Center, IA: Dordt College Press, 2005.

Searle, John R. "Minds, Brains and Programs." *Behavioral and Brain Sciences* 3, no. 3 (1980): 417-57.

Seerveld, Calvin G. *A Turnabout in Aesthetics to Understanding*. Toronto: Institute for Christian Studies, 1974.

Sharkey, Amanda, and Noel Sharkey. "Children, the Elderly, and Interactive Robots." *Robotics and Automation Magazine, IEEE* 18, no. 1 (March 2011): 32-38.

Sharkey, Noel. "Automated Killers and the Computing Profession." *Computer* 40, no. 11 (2007): 124-23.

Shasha, Dennis, and Cathy Lazere. *Out of Their Minds: the Lives and Discoveries of 15 Great Computer Scientists.* New York: Copernicus, 1998.

Singer, Peter W. *Wired for War: The Robotics Revolution and Conflict in the 21st Century.* New York: Penguin, 2009.

Small, Gary. *iBrain: Surviving the Technological Alteration of the Modern Mind.* New York: William Morrow, 2008.

Smedes, Lewis. *My God and I.* Grand Rapids: Eerdmans, 2003.

Smith, Cynthia E. *Design for the Other 90%.* New York: Cooper-Hewitt National Design Museum, 2007.

Smith, James K. A. *Desiring the Kingdom.* Grand Rapids: Baker Academic, 2009.

Spykman, Gordon J. *Reformational Theology: A New Paradigm for Doing Dogmatics.* Grand Rapids: Eerdmans, 1992.

Standage, Tom. *The Victorian Internet.* New York: Walker & Company, 2007.

Sundem, Garth. *Geek Logik: 50 Foolproof Equations for Everyday Life.* New York: Workman Publishing Company, 2006.

Taylor, Alexander L., III, Michael Moritz and Peter Stoler. "The Wizard Inside the Machine." *Time,* April 1984, pp. 56-63.

Tesla, Nikola. "The Wonder World to be Created by Electricity." *Manufacturer's Record* (September 1915).

Thoreau, Henry David. *Walden.* Princeton, NJ: Princeton University Press, 1973.

Torvalds, Linus, and David Diamond. *Just for Fun: The Story of an Accidental Revolutionary.* New York: HarperCollins, 2001.

Turing, Alan. "Computing Machinery and Intelligence." *Mind* 59 (1950): 433-60.

Turkle, Sherry. *Alone Together: Why We Expect More from Technology and Less from Each Other.* New York: Basic Books, 2011.

Venkat, Kumar. "Delving into the Digital Divide." *IEEE Spectrum* 39, no. 2 (February 2002): 14-16.

Vinge, Vernor. "The Coming Technological Singularity." Presented at the VISION-21 Symposium, sponsored by NASA Lewis Research Center and the Ohio Aerospace Institute, March 30-31, 1993.

Walsh, Brian J., and J. Richard Middleton. *The Transforming Vision: Shaping a Christian World View.* Downers Grove, IL: InterVarsity Press, 1984.

Weiss, Lora G. "Autonomous Robots in the Fog of War." *IEEE Spectrum* 48, no. 8 (August 2011): 31-34, 56-57.

Weizenbaum, Joseph. *Computer Power and Human Reason: From Judgment to Calculation.* New York: W. H. Freeman, 1976.

Wenham, Gordon J. *Genesis 1-15.* Word Biblical Commentary 1. Waco, TX: Word Books, 1987.

White, Lynn. "The Historical Roots of Our Ecological Crisis." *Science* 155, no. 3767 (March 1967): 1203-7.

Wilson, Daniel H. *How To Survive a Robot Uprising: Tips on Defending Yourself Against the Coming Rebellion.* New York: Bloomsbury USA, 2005.

Wolters, Albert M. *Creation Regained.* Grand Rapids: Eerdmans, 1985.

———. "Living the Future Now (1)." *Christian Educators Journal* 39, no. 1 (October 1999): 4-7.

———. "Living the Future Now (2)." *Christian Educators Journal* 39, no. 2 (December 1999): 16-19.

———. "Worldview and Textual Criticism in 2 Peter 3:10." *Westminster Theological Journal* 49 (1987): 405-13.

Wolterstorff, Nicholas. *Art in Action: Toward a Christian Aesthetic.* Grand Rapids: Eerdmans, 1980.

———. "On Christian Learning." In *Stained Glass: Worldviews and Social Science,* edited by Paul A. Marshall, Sander Griffioen and Richard J. Mouw, pp. 56-80. Lanham, MD: University Press of America, 1989.

———. *Reason Within the Bounds of Religion.* Grand Rapids: Eerdmans, 1999.

———. *Until Justice and Peace Embrace.* Grand Rapids: Eerdmans, 1983.

Yourdon, Edward. *Death March: The Complete Software Developer's Guide to Surviving 'Mission Impossible' Projects.* Upper Saddle River, NJ: Prentice Hall, 1999.

Zorpette, Glenn. "Waiting for the Rapture." *IEEE Spectrum* 45, no. 6 (June 2008): 32-35.

Author and Subject Index